AF777369

Materials for Sustainable Development —Challenges and Opportunities

MATERIALS RESEARCH SOCIETY
SYMPOSIUM PROCEEDINGS VOLUME 1492

Materials for Sustainable Development—Challenges and Opportunities

Symposium held November 25–30, 2012, Boston, Massachusetts, U.S.A.

EDITORS

Alex King
Critical Materials Institute
Ames, Iowa, U.S.A.

Laura Espinal
National Institute of Standards and Technology
Gaithersburg, Maryland, U.S.A.

John Poate
Colorado School of Mines
Golden, Colorado, U.S.A.

Enrico Traversa
King Abdullah University of Science and Technology (KAUST)
Thuwal, Saudi Arabia

Mary M. Poulton
The University of Arizona
Tucson, Arizona, U.S.A.

Samuel S. Mao
Lawrence Berkeley National Laboratory
Berkeley, California, U.S.A.

Steven Duclos
GE Global Research
Niskayuna, New York, U.S.A.

Marie-Isabelle Baraton
Centre Europeen de la Ceramique
Limoges, France

MRS

Materials Research Society
Warrendale, Pennsylvania

CAMBRIDGE
UNIVERSITY PRESS

CAMBRIDGE UNIVERSITY PRESS
Cambridge, New York, Melbourne, Madrid, Cape Town,
Singapore, São Paulo, Delhi, Mexico City

Cambridge University Press
32 Avenue of the Americas, New York, NY 10013-2473, USA

www.cambridge.org
Information on this title: www.cambridge.org/9781605114699

Materials Research Society
506 Keystone Drive, Warrendale, PA 15086
http://www.mrs.org

© Materials Research Society 2013

This publication is in copyright. Subject to statutory exception and to the provisions of relevant collective licensing agreements, no reproduction of any part may take place without the written permission of Cambridge University Press.

This book has been registered with Copyright Clearance Center, Inc. For further information please contact the Copyright Clearance Center, Salem, Massachusetts.

First published 2013

This work was performed under the sponsorship of the U.S. Department of Commerce, National Institute of Standards and Technology Grant No. 60NANB12D221.

CODEN: MRSPDH

ISBN: 978-1-60511-469-9 Hardback

Cambridge University Press has no responsibility for the persistence or accuracy of URLs for external or third-party Internet Web sites referred to in this publication and does not guarantee that any content on such Web sites is, or will remain, accurate or appropriate.

CONTENTS

*Invited Paper

MATERIAL RESOURCES

SMART MATERIALS

*Invited Paper

ENVIRONMENT AND HEALTH

EFFICIENCY AND FUEL SWITCHING

PREFACE

Symposium D, "Energy Critical Materials" and Symposium G, "Materials as Tools for Sustainability," were held on November 25–30 at the 2012 MRS Fall Meeting in Boston, Massachusetts.

Many technologies that currently impart significant benefits to our society cannot continue indefinitely without depleting or despoiling key resources, and more sustainable paths must be sought. Sustainable development that raises global standards of living and promotes economic growth calls for materials science to develop advanced technologies and strategies to use our finite material and energy resources more efficiently to minimize impact on environment and human health. As an example, both clean and traditional energy technologies have critical dependencies on specific elements. Therefore, challenges in the supply of these critical materials could directly affect the adoption of new technologies and ultimately threaten energy security. This symposium proceedings volume presents a compilation of emerging research on sustainable development in the context of product manufacturing, catalysis, transportation, and energy-critical materials. The papers are divided into six sections: (1) Energy Storage, (2) Rare Earths and Magnets, (3) Materials Resources, (4) Smart Materials, (5) Environment and Health, (6) Efficiency and Fuel Switching. Each paper in this volume provides a glimpse of the exciting recent developments occurring in several aspects of sustainability, including: critical elements for next-generation batteries, phase change materials for energy efficient buildings, weathering resistant materials, 3-D printing technologies, flame retardants with reduced toxicity, and stimulating life-cycle analysis examples in various applications. We hope these papers convey the breadth of exciting advancements happening in the area of sustainability including critical elements for energy.

Alex King
John Poate
Mary M. Poulton
Steven Duclos
Laura Espinal
Enrico Traversa
Samuel S. Mao
Marie-Isabelle Baraton

June 2013

ACKNOWLEDGMENTS

The papers published in this volume result from two MRS Fall 2012 symposia—D and G. We sincerely thank all of the oral and poster presenters of the symposia who contributed to this proceedings volume. We also thank the reviewers of these manuscripts, who provided valuable feedback to the editors and to the authors. It is an understatement to say that the symposia and the proceedings would not have happened without the organizational help of the Materials Research Society and its staff, particularly the publications staff for guiding us smoothly through the submission/review process and constantly nudging us to move forward. The organizers of Symposium G thank the National Institute of Standards and Technology for its financial support.

Symposium support was received from:

The Ames Laboratory (D)
The University of Arizona (D)
Colorado School of Mines (D)
GE Global Research Center (D)
Sigma-Aldrich, Inc. (D)
National Institute of Standards and Technology (G)

MATERIALS RESEARCH SOCIETY SYMPOSIUM PROCEEDINGS

Volume 1477 — Low-Dimensional Bismuth-based Materials, 2012, S. Muhl, R. Serna, A. Zeinert, S. Hirsekor, ISBN 978-1-60511-454-5
Volume 1478 — Nanostructured Carbon Materials for MEMS/NEMS and Nanoelectronics, 2012, A.V. Sumant, A.A. Balandin, S.A. Getty, F. Piazza, ISBN 978-1-60511-455-2
Volume 1479 — Nanostructured Materials and Nanotechnology—2012, 2012, C. Gutiérrez-Wing, J.L. Rodríguez-López, O. Graeve, M. Munoz-Navia, ISBN 978-1-60511-456-9
Volume 1480 — Novel Characterization Methods for Biological Systems, 2012, P.S. Bermudez, J. Majewski, N. Alcantar, A.J. Hurd, ISBN 978-1-60511-457-6
Volume 1481 — Structural and Chemical Characterization of Metals, Alloys and Compounds—2012, 2012, A. Contreras Cuevas, R. Pérez Campos, R. Esparza Muñoz, ISBN 978-1-60511-458-3
Volume 1482 — Photocatalytic and Photoelectrochemical Nanomaterials for Sustainable Energy, 2012, L. Guo, S.S. Mao, G. Lu, ISBN 978-1-60511-459-0
Volume 1483 — New Trends in Polymer Chemistry and Characterization, 2012, L. Fomina, M.P. Carreón Castro, G. Cedillo Valverde, J. Godínez Sánchez, ISBN 978-1-60511-460-6
Volume 1484 — Advances in Computational Materials Science, 2012, E. Martínez Guerra, J.U. Reveles, A. Aguayo González, ISBN 978-1-60511-461-3
Volume 1485 — Advanced Structural Materials—2012, 2012, H. Calderon, H.A. Balmori, A. Salinas, ISBN 978-1-60511-462-0
Volume 1486E — Nanotechnology-enhanced Biomaterials and Biomedical Devices, 2012, L. Yang, M. Su, D. Cortes, Y. Li, ISBN 978-1-60511-463-7
Volume 1487E — Biomaterials for Medical Applications—2012, 2012, S. Rodil, A. Almaguer, K. Anselme, J. Castro, ISBN 978-1-60511-464-4
Volume 1488E — Concrete with Smart Additives and Supplementary Cementitious Materials, 2012, L.E. Rendon Diaz Miron, B. Martinez Sanchez, K. Kovler, N. De Belie, ISBN 978-1-60511-465-1
Volume 1489E — Compliant Energy Sources, 2013, D. Mitlin, ISBN 978-1-60511-466-8
Volume 1490 — Thermoelectric Materials Research and Device Development for Power Conversion and Refrigeration, 2013, G.S. Nolas, Y. Grin, A. Thompson, D. Johnson, ISBN 978-1-60511-467-5
Volume 1491E — Electrocatalysis and Interfacial Electrochemistry for Energy Conversion and Storage, 2013, T.J. Schmidt, V. Stamenkovic, M. Arenz, S. Mitsushima, ISBN 978-1-60511-468-2
Volume 1492 — Materials for Sustainable Development—Challenges and Opportunities, 2013, M-I. Baraton, S. Duclos, L. Espinal, A. King, S.S. Mao, J. Poate, M.M. Poulton, E. Traversa, ISBN 978-1-60511-469-9
Volume 1493 — Photovoltaic Technologies, Devices and Systems Based on Inorganic Materials, Small Organic Molecules and Hybrids, 2013, K.A. Sablon, J. Heier, S.R. Tatavarti, L. Fu, F.A. Nüesch, C.J. Brabec, B. Kippelen, Z. Wang, D.C. Olson, ISBN 978-1-60511-470-5
Volume 1494 — Oxide Semiconductors and Thin Films, 2013, A. Schleife, M. Allen, S.M. Durbin, T. Veal, C.W. Schneider, C.B.Arnold, N. Pryds, ISBN 978-1-60511-471-2
Volume 1495E — Functional Materials for Solid Oxide Fuel Cells, 2013, J.A. Kilner, J. Janek, B. Yildiz, T. Ishihara, ISBN 978-1-60511-472-9
Volume 1496E — Materials Aspects of Advanced Lithium Batteries, 2013, V. Thangadurai, ISBN 978-1-60511-473-6
Volume 1497E — Hierarchically Structured Materials for Energy Conversion and Storage, 2013, P.V. Braun, ISBN 978-1-60511-474-3
Volume 1498 — Biomimetic, Bio-inspired and Self-Assembled Materials for Engineered Surfaces and Applications, 2013, M.L. Oyen, S.R. Peyton, G.E. Stein, ISBN 978-1-60511-475-0
Volume 1499E — Precision Polymer Materials—Fabricating Functional Assemblies, Surfaces, Interfaces and Devices, 2013, C. Hire, ISBN 978-1-60511-476-7
Volume 1500E — Next-Generation Polymer-Based Organic Photovoltaics, 2013, M.D. Barnes, ISBN 978-1-60511-477-4
Volume 1501E — Single-Crystalline Organic and Polymer Semiconductors—Fundamentals and Devices, 2013, S.R. Parkin, ISBN 978-1-60511-478-1
Volume 1502E — Membrane Material Platforms and Concepts for Energy, Environment and Medical Applications, 2013, B. Hinds, F. Fornasiero, P. Miele, M. Kozlov, ISBN 978-1-60511-479-8
Volume 1503E — Colloidal Crystals, Quasicrystals, Assemblies, Jammings and Packings, 2013, S.C. Glotzer, F. Stellacci, A. Tkachenko, ISBN 978-1-60511-480-4
Volume 1504E — Geometry and Topology of Biomolecular and Functional Nanomaterials, 2013, A. Saxena, S. Gupta, R. Lipowsky, S.T. Hyde, ISBN 978-1-60511-481-1

MATERIALS RESEARCH SOCIETY SYMPOSIUM PROCEEDINGS

Volume 1505E — Carbon Nanomaterials, 2013, J.J. Boeckl, W. Choi, K.K.K. Koziol, Y.H. Lee, W.J. Ready, ISBN 78-1-60511-482-8
Volume 1506E — Combustion Synthesis of Functional Nanomaterials, 2013, R.L. Vander Wal, ISBN 978-1-60511-483-5
Volume 1507E — Oxide Nanoelectronics and Multifunctional Dielectrics, 2013, P. Maksymovych, J.M. Rondinelli, A. Weidenkaff, C-H. Yang, ISBN 978-1-60511-484-2
Volume 1508E — Recent Advances in Optical, Acoustic and Other Emerging Metamaterials, 2013, K. Bertoldi, N. Fang, D. Neshev, R. Oulton, ISBN 978-1-60511-485-9
Volume 1509E — Optically Active Nanostructures, 2013, M. Moskovits, ISBN 978-1-60511-486-6
Volume 1510E — Group IV Semiconductor Nanostructures and Applications, 2013, L. Dal Negro, C. Bonafos, P. Fauchet, S. Fukatsu, T. van Buuren, ISBN 978-1-60511-487-3
Volume 1511E — Diamond Electronics and Biotechnology—Fundamentals to Applications VI, 2013, Y. Zhou, ISBN 978-1-60511-488-0
Volume 1512E — Semiconductor Nanowires—Optical and Electronic Characterization and Applications, 2013, J. Arbiol, P.S. Lee, J. Piqueras, D.J. Sirbuly, ISBN 978-1-60511-489-7
Volume 1513E — Mechanical Behavior of Metallic Nanostructured Materials, 2013, Q.Z. Li, D. Farkas, P.K. Liaw, B. Boyce, J.Wang, ISBN 978-1-60511-490-3
Volume 1514 — Advances in Materials for Nuclear Energy, 2013, C.S. Deo, G. Baldinozzi, M.J. Caturla, C-C. Fu, K. Yasuda, Y. Zhang, ISBN 978-1-60511-491-0
Volume 1515E — Atomic Structure and Chemistry of Domain Interfaces and Grain Boundaries, 2013, S.B. Sinnott, B.P. Uberuaga, E.C. Dickey, R.A. De Souza, ISBN 978-1-60511-492-7
Volume 1516 — Intermetallic-Based Alloys—Science, Technology and Applications, 2013, I. Baker, S. Kumar, M. Heilmaier, K. Yoshimi, ISBN 978-1-60511-493-4
Volume 1517 — Complex Metallic Alloys, 2013, M. Feuerbacher, Y. Ishii, C. Jenks, V. Fournée, ISBN 978-1-60511-494-1
Volume 1518 — Scientific Basis for Nuclear Waste Management XXXVI, 2012, N. Hyatt, K.M. Fox, K. Idemitsu, C. Poinssot, K.R. Whittle, ISBN 978-1-60511-495-8
Volume 1519E — Materials under Extreme Environments, 2013, R.E. Rudd, ISBN 978-1-60511-496-5
Volume 1520E — Structure-Property Relations in Amorphous Solids, 2013, Y. Shi, M.J. Demkowicz, A.L. Greer, D. Louca, ISBN 978-1-60511-497-2
Volume 1521E — Properties, Processing and Applications of Reactive Materials, 2013, E. Dreizin, ISBN 978-1-60511-498-9
Volume 1522E — Frontiers of Chemical Imaging—Integrating Electrons, Photons and Ions, 2013, C.M. Wang, J.Y. Howe, A. Braun, J.G. Zhou, ISBN 978-1-60511-499-6
Volume 1523E — Materials Informatics, 2013, R. Ramprasad, R. Devanathan, C. Breneman, A. Tkatchenko, ISBN 978-1-60511-500-9
Volume 1524E — Advanced Multiscale Materials Simulation—Toward Inverse Materials Computation, 2013, D. Porter, ISBN 978-1-60511-501-6
Volume 1525E — Quantitative *In-Situ* Electron Microscopy, 2013, N.D. Browning, ISBN 978-1-60511-502-3
Volume 1526 — Defects and Microstructure Complexity in Materials, 2013, A. El-Azab, A. Caro, F. Gao, T. Yoshiie, P. Derlet, ISBN 978-1-60511-503-0
Volume 1527E — Scanning Probe Microscopy—Frontiers in Nanotechnology, 2013, M. Rafailovich, ISBN 978-1-60511-504-7
Volume 1528E — Advanced Materials Exploration with Neutrons and Synchrotron X-Rays, 2013, J.D. Brock, ISBN 978-1-60511-505-4
Volume 1529E — Roll-to-Roll Processing of Electronics and Advanced Functionalities, 2013, T. Blaudeck, G. Cho, M.R. Dokmeci, A.B. Kaul, M.D. Poliks, ISBN 978-1-60511-506-1
Volume 1530E — Materials and Concepts for Biomedical Sensing, 2013, P. Kiesel, M. Zillmann, H. Schmidt, B. Hutchison, ISBN 978-1-60511-507-8
Volume 1531E — Low-Voltage Electron Microscopy and Spectroscopy for Materials Characterization, 2013, R.F. Egerton, ISBN 978-1-60511-508-5
Volume 1532E — Communicating Social Relevancy in Materials Science and Engineering Education, 2013, K. Chen, R. Nanjundaswamy, A. Ramirez, ISBN 978-1-60511-509-2
Volume 1533E — The Business of Nanotechnology IV, 2013, L. Merhari, D. Cruikshank, K. Derbyshire, J. Wang, ISBN 978-1-60511-510-8

MATERIALS RESEARCH SOCIETY SYMPOSIUM PROCEEDINGS

Volume 1534E — Low-Dimensional Semiconductor Structures, 2012, T. Torchyn, Y. Vorobie, Z. Horvath, ISBN 978-1-60511-511-5

Prior Materials Research Society Symposium Proceedings available by contacting Materials Research Society

Energy Storage

Mater. Res. Soc. Symp. Proc. Vol. 1492 © 2013 Materials Research Society
DOI: 10.1557/opl.2013.55

Lithium - the Metal of the Future?

Ihor A. Kunasz[1]

[1]TRU Group, 95 Prince Arthur Avenue, Suite 117, Toronto, ON, Canada

ABSTRACT

The second half of the twentieth century saw a dramatic shift in lithium chemicals production from traditional pegmatite sources to brines. Today, the bulk of lithium carbonate, which serves as the raw material for various downstream lithium chemicals, including lithium metal for the lithium batteries, is produced from the brines of the Salar de Atacama, Chile, the Salar del Hombre Muerto, Argentina and Clayton Valley Nevada, U.S.A. There is minor production in Tibet and the People's Republic of China (PRC). Australian spodumene concentrates are converted to lithium carbonate in the PRC.

The resurgence in the potential development of electric cars has resulted in the increased exploration for and identification of potential new lithium brine operations and the reassessment of some pegmatite deposits.

A number of predictions for a potentially large electric car market scenario have raised questions on the availability of sufficient lithium resources. However, since the original 1976 report on global lithium resources by the National Academies of Sciences and Engineering, newly identified deposits have almost quadrupled the total potentially available lithium resources. Based on the best predictions, the lithium supply is more than adequate to meet the demand for electric cars well into the 21st century.

INTRODUCTION

There have been three significant mileposts in the history of lithium minerals and chemicals production. Prior to WWII, small quantities of lepidolite mica from the Stewart mine, San Diego County, California were used in the production of alkaline storage batteries. During WWII, lithium hydroxide was used for rescue apparatus and lithium greases. The first big boost in lithium mining came from the Atomic Energy Commission program to develop the hydrogen bomb in the 1950's. Following the closure of the program, lithium minerals and chemicals were used in ceramic and glass applications [1]

The potential development of fusion and the development of electric automobiles in the 1970's prompted the U.S. government to undertake an extensive brine exploration program to increase the lithium resource base in the United States. However, the program was not successful and the predicted production of a large number of electric vehicles did not materialize.

Today, the renewed interest in the potential of lithium use in electric cars and energy storage systems, largely driven by government incentives for the development of green, pollution-free energy resulted in a resurgence of exploration for and potential development of new lithium resources, both brines and pegmatites. However, after several years of research and development of batteries for the electric car, the jury is still out on the future of these endeavors.

LITHIUM SOURCES

Pegmatites

Until the 1960's, the traditional sources of lithium mineral and chemicals have been pegmatites. These are residual late-phase post-granitic formations. They may be zoned or un-zoned. They constitute the rare element deposits and may contain cassiterite, columbite-tantalite, wolframite, beryl and various lithium minerals (spodumene, amblygonite, petalite, lepidolite).

The most important lithium mineral has been and is spodumene, a lithium alumino-silicate (LiAlSiO3), and used predominantly in ceramic applications.

Petalite ($LiAlSi_4O_{10}$) is also a preferred lithium mineral in ceramic application where low iron is needed to produce very white ceramics.

In the 1950s, lepidolite mica from Zimbabwe (formerly Southern Rhodesia) was imported to the US and converted to lithium hydroxide from which the Li^6 isotope was recovered to produce the hydrogen (tritium) bomb.

The best know and most productive of the zoned pegmatites operations was the Etta mine, in the Black Hills of South Dakota, where spodumene was produced for over 50 years. Several other pegmatites in California, northern New Mexico and Colorado has noteworthy records of production in those early years (1).

From the 1950's until the mid-1990, Foote Mineral Company and Lithium Corporation of America recovered spodumene from pegmatites associated with the regional Cherryville quartz-monzonite, in North Carolina. Both companies produced a chemical grade concentrate for conversion into downstream lithium chemicals. Foote Mineral Company also produced a low-iron concentrate for application in the "Corningware" pyro-ceramic. However, following the production of lithium from brines by SQM in 1997, both companies closed their North Carolina operations.

Today, by far the most important producer of lithium concentrates Talison Lithium Limited. Talison mines a spodumene pegmatite in Greenbushes, Australia, where it operates two recovery circuits: a high-grade concentrate for the ceramic and glass industries and a chemical grade for conversion to lithium carbonate. The company has announced an engineering study for building a 20,000 metric tons lithium carbonate plant in Western Australia [2]. The Mt. Cattlin deposit is a high iron type. At full capacity Galaxy Resources will process 137,000 metric tons of spodumene concentrates to feed its lithium carbonate plant in Jiangsu, People's Republic of China (PRC), expected to produce 17,000 metric tons of lithium carbonate [3].

Active mining, albeit small. operations occur in the PRC, Brazil, Portugal, and Zimbabwe [17].

Brines

Lithium and other elements occur in nearly saturated sodium chloride brines containing various subordinate amounts of lithium, potassium, magnesium, sulfate, and boron.

Figure 1 illustrates the principal types of evaporitic basin which contain lithium brines:

GENERAL CHARACTERISTICS OF LITHIUM BRINE DEPOSITS-BEARING SALINE BASINS

- They occur within Tertiary or Recent volcanic belts.
- They are closed structural depressions.
- They occur within the desert areas of the earth.

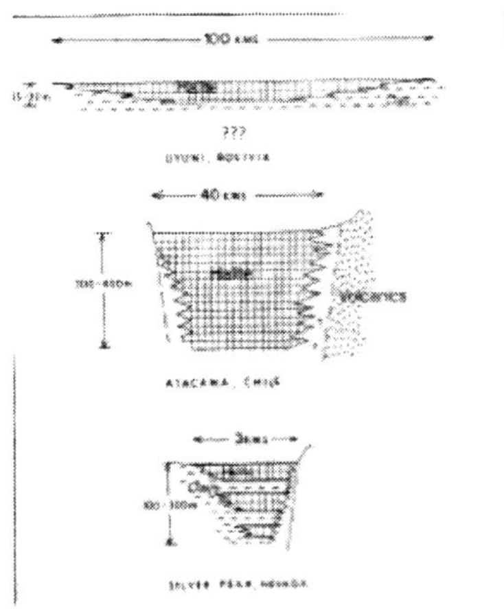

Figure 1 - Types of known continental lithium brine deposits

An evaluation of the many salars identified in the various regions of the world revealed that brine deposits differ in compositional character, both in lithium concentration and in the distribution of various ionic ratios [table 1]. Consequently, the various ionic ratios require different processing approaches to prevent losses of lithium due to the precipitation of complex salts, and economically recover lithium and associated salts.

COMPARATIVE BRINES CHEMISTRIES (%)

Brine Deposit	Li	K	SO4	Mg	SO4/Li	SO4/K	SO4/Mg	Mg/Li
SALAR DE ATACAMA	0.15	1.85	1.46	0.96	9.7	0.79	1.52	6.4
HOMBRE MUERTO	0.07	0.6	0.9	0.21	12.9	0.67	4.29	3.0
SILVER PEAK NEVADA	0.02	0.8	0.71	0.03	35.5	1.13	23.67	1.5
RINCON	0.033	0.62	1.1	0.28	33.3	0.56	3.93	8.5
OLAROZ	0.066	0.55	1.7	0.19	25.8	0.32	8.95	2.9
CAUCHARI	0.051	0.42	1.57	0.15	30.8	0.27	10.83	2.8
DIABLILLOS	0.05	0.21	0.75	0.55	15.0	0.28	1.36	11.0
UYUNI	0.045	1.2	1.4	1.07	31.1	0.86	1.31	23.8

TRU Group

Table I - Chemical composition of various continental brine deposits

The most critical aspect is to eliminate the various impurities (principally magnesium and sulfate) during the evaporative process and keep the lithium in solution until the end of the solar concentration cycle.

Brine Deposits

Clayton Valley, Nevada, United States

In the 1960's, following the WWII efforts to locate critical minerals, potassium was identified in the brines of Clayton Valley, Nevada. Subsequent chemical analyses also identified substantial concentrations of lithium in the brines (200-300 ppm Li). Following the acquisition of the leases by Foote Mineral Company (Chemetall), Dr. Wayne Barrett, a physical chemist and vice-president of research, solved the process for the recovery of lithium from the chemically complex brine [4] In the late 1960's, the deposit became the first successful source of lithium carbonate from a brine; it is still producing today.

Salar de Atacama, Chile

In 1970, Chilean Geologists [5] identified high lithium concentrations (1500 ppm Li) in the brines of Salar de Atacama in Northern Chile. Subsequently, the Chilean government development agency, CORFO, undertook an extensive exploration program in northern Chile, and identified lithium in a number of additional salares (Punta Negra, Pedernales, Maricunga, and others). Today the Salar de Atacama is the largest producer of lithium carbonate in the world.

Salar de Hombre Muerto, Argentina

In the1980's, Argentinean geologists conducted an extensive survey of the salares of the Andean Puna [6] and confirmed high lithium concentrations. One of these, the Salar del Hombre Muerto (700 ppm Li) was put into production by FMC Corporation. The Salar del Rincon (Sentient Corp) is producing pilot plant quantities of lithium carbonate. To-date, as a result of the renewed interest in lithium for electric cars, five additional salars are being evaluated for their potential production of lithium carbonate.

Salar de Uyuni, Bolivia

In1976, the Servicio Geologico de Bolivia (GEOBOL) and the United States Geological Survey (USGS) identified lithium (250- 450 ppm Li) in upper layers of the largest salar in the world, the Salar de Uyuni, which covers an area of nearly 10,000 km^2. Subsequently, ORSTOM, in conjunction with GEOBOL, delineated a very substantial lithium resource in the upper 20 meters of the salt layer [7]. Later drilling confirmed that the salt layer was thicker than 200 meters, suggesting a much larger lithium reserve. Preliminary efforts at producing lithium chemicals from this complex brine by the Bolivian government have so far met with little success.

Qinghai Basin, People's Republic of China and Lake Zabuye, Tibet

Two areas are producing lithium carbonate from brines. However, technical chemistry problems have resulted in production issues and significantly lower production volume than originally announced:

- Qinghai basin (Lakes Dabusan and Tajinaer) in Northern China and
- Lake Zabuye in western Tibet

Other potential sources

Continental Brines – Argentina

In the past few years, several companies have explored a number of continental brine deposits in the Argentinean Altiplano. The salar de Rincon is producing pilot quantities of lithium carbonate. The salares of Olaroz, Cauchari, Sal de Vida, Diablillos, and others contain a few to several hundred ppm Li and could potentially be brought into production, if a high demand scenario for lithium develops. The low concentrations brines will likely not be competitive with existing high lithium brines. All must be gauged against the Salar de Atacama where the lithium concentration is three to four times higher.

Pegmatites- Canada

A great number of pegmatites were identified in Canada in the 1950's. Recently, several companies have evaluated the merits of redeveloping those resources, which have lithium concentrations ranging form 0.5 – 0.7% Li [17].

Clays
Lithium (0.6% Li) occurs in hectorite, swelling montmorillonite clay, produced by the alteration of volcanic glass by geothermal waters. One such classical deposit is mined in Hector, CA and used in cosmetics.
In Nevada, Western Mining Corp. is evaluating a large resource of a low-grade lithium clay deposit (0.25- 0.3% Li) at the Kings Valley Property [8].
Geothermal Brines
Lithium (200 ppm Li) has long been identified in the geothermal brines of the Imperial Valley, CA. Recently, Simbol Material Corp. has recently announced the success in recovering lithium and other metals from the flashed brine. If successful, the operation could supply large quantities of lithium carbonate [9].
Oil Field Brines
Lithium (200-300 ppm Li) has long been known to occur in the brines associated with oil fields in an area extending across several states from Louisiana to Texas. The brines are processed fro their bromine content. Albemarle Corp. has recently announced a technology to recover lithium from their spent brine. [10]
Jadarite
Jadarite is a complex silicate mineral of lithium and boron. Rio Tinto outlined a substantial resource of lithium and boron in a drilling program in the Jadar Valley of Serbia. The company has completed a conceptual study to produce lithium and boron. [11]

LITHIUM PRODUCERS

Continental Brines
The most significant structural change resulted from the discovery of lithium at the Salar de Atacama. Chile, and the Salar del Hombre Muerto in Argentina.

In Chile, two companies process the brine for lithium and potassium:

- *Sociedad Chilena de Litio, Ltd.* (SCL-CHEMTALL), a subsidiary of Rockwood Holdings, operates a complex evaporation pond system in the southern part of the Salar de Atacama

- *Sociedad Quimica y Minera de Chile, S. A.* (SQM). Extracts brine from the northern part of the salar, which it processes in a dual pond system.

In Argentina, one company operates a brine deposit:

- *Lithium Corporation of America (Minera del Altiplano),* part of the FMC group, produces lithium from the Salar del Hombre Muerto and

These three companies account for 79% of the lithium carbonate production in the world [12].
Sentient Corp. produces minor amounts of lithium carbonate at the Salar de Rincon.

Rockwood Holdings

Through its subsidiary, CHEMETALL - SCL, the company produces lithium chemicals from two brine deposits:

- The southern portion of the Salar de Atacama (Sociedad Chilena de Litio, Ltd.)
- Clayton Valley, Nevada

SCL pumps high lithium brines into large solar evaporation ponds, where it concentrates where it controls the phase chemistry. The salar brine is essentially a saturated sodium chloride brine, with various amounts of potassium, magnesium, sulfate and boron. The complex chemistry was solved by the mixing of a calcium-rich brine with the dominant sulfate-rich brine to precipitate gypsum and create a chloride phase that allowed the concentration of a 6% lithium chloride final brine.

Figure 2 - Salar de Atacama, Chile, pond evaporation complex of SCL

The concentrated brine is trucked to the chemical production facilities at La Negra, located east of the port City of Antofagasta. It also produces lithium chemicals from its Clayton Valley NV operation.

In February of 2012, Rockwood Holdings announced in that it plans to invest $140 to build an additional 20,000 metric ton lithium carbonate plant at La Negra. The company expects this plant will be on stream by the end of 2013. The new investment, along with the company's current $75 million expansion program of its lithium production in the United States, which includes expansion of its brine pond system at Silver Peak, NV, and construction of a battery grade lithium hydroxide plant and a global technical center at Kings Mountain, NC, will increase the total annual production capacity to 50,000 metric tons of lithium carbonate equivalent by end of 2013 [13]

Rockwood Holdings has recently acquired Talison, the largest global producer of spodumene concentrates. Low-iron spodumene is used directly in ceramic and glass applications in competition to lithium carbonate. This acquisition makes Rockwood Holdings the most important player in the consolidated lithium industry.

Sociedad Quimica y Minera

The Chilean company is the largest producer of lithium carbonate and one of the largest lithium hydroxide in the world. It operates two large pond evaporation complexes in the northern portion of the Salar de Atacama (Figure 3). SQM recovers two types of brine : a high-potassium high-lithium brine to precipitate potassium chloride for the production of potassium nitrate and a mixed high-sulfate high-potassium brine to produce potassium sulfate.

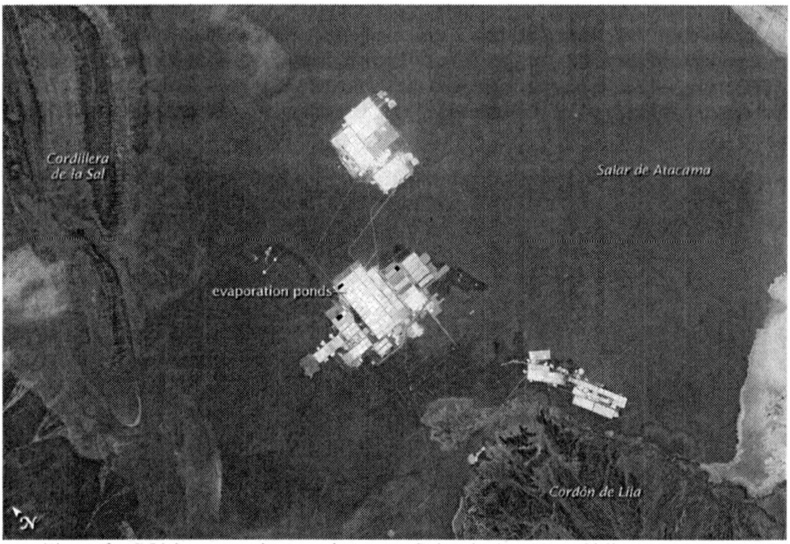

Figure 3 - SQM evaporation pond system, Salar de Atacama, Chile

It is critical to understand that SQM has a unique strength. The brine volume pumped to produce the potash required for fertilizer requirements far exceeds the lithium necessary for the lithium chemical production capacity. A portion of the final concentrated brine is shipped to the chemical processing plants at the Salar del Carmen near Antofagasta, and the excess enriched lithium brine is pumped back into the porous salt of the salar. Of critical importance from a lithium resource perspective is that this effectively creates a future concentrated secondary lithium reserve ready for low cost exploitation should the market rapidly expand. In 2011, SQM reported an annual lithium carbonate capacity of 48,000 metric tons lithium carbonate equivalents [12].

FMC – Argentina

Following the closure of their Bessemer, N. C. pegmatite operation in 1997, FMC undertook the development of the brine deposit at the Salar del Hombre Muerto, in the Argentinean Altiplano. FMC extracts lithium from the brines by a proprietary ion exchange method and ships the purified lithium chloride brine to its plant facilities near Salta Argentina. FMC plans an expansion of 30% to 23,000 metric tons lithium carbonate equivalents). FMC is the second largest producers of downstream lithium chemicals [14]

LITHIUM RESOURCES

There have been a number of different resource figures prepared by a variety of authors. Unfortunately, many of these are secondary sources and authors have used considerable editorial liberties and conclusions. The problem with these authors is that they are not familiar with the specific geological/operational conditions of the various deposits. They are generalists, never having conducted exploration for nor operated either lithium pegmatite or brine deposits. The lithium specialists are few. I have quoted the reliable source of Keith Evans, who has in fact done both [15]. The author of this paper has a long professional history of d management of lithium properties as chief geologist for Foote Mineral company, has developed prediction models for brine deposits and has evaluated most of the know lithium deposits of the world [16]. He presently works with a highly reputable group of lithium specialists at TRU Group, which includes experts from resources through processing and metal production and has a network of specialists who cover the global lithium industry. TRU has also conducted detailed first hand evaluations of lithium properties from resources through feasibility across the world.

In 1976, the National Academy of Sciences and Engineering formed a National Research Council Committee on Nuclear and Alternative Energy Systems (CONEAS) to report on the role of nuclear power in the context of alternative energy systems in the time period 1985 to 2010. CONEAS was organized into four main panels and twenty-six sub panels and the Lithium Sub Panel was tasked to report on lithium raw material availability. This group was chaired by Dr. Thomas Kesler, formerly with the USGS and the leading authority on the North Carolina tin-spodumene belt the, then, dominant source of lithium, Dr. James Vine of the USGS and the head of its Lithium Resource Appraisal Group, Dr. Ihor Kunasz of the Foote Mineral Company and Keith Evans representing Lithium Corporation of America. The lithium resources tonnages estimated in the panel report stood at 10.65 million tonnes [15]. By 2010, Evans updated the data and reported a total lithium resource estimate of 34.5 million metric tons Li [16]. To this figure must be added the newly reported resources in Argentina for Lithium Americas at the Salar de Cauchari (921,000 tons Li), for Rodinia at Salar De Diablillos (930,000 metric tons Li), and for Galaxy Resources at Sal de Vida (1,359,000metric tons Li), which together have added resources of 3,200,000 metric tons of Li, increasing the global resource base by almost 10%. The most recent additions increase the lithium resource figure to new total of nearly 39 million tons Li. It behooves the reader to realize that in a very short period of time, the perceived increase in lithium demand has resulted in a flurry of exploration projects, which are a portent of the future trend in lithium discoveries. A detailed review of the various companies involved in lithium exploration and potential development has been recently published by Clarke [18]. As was the case in the 1970 at the Corning Lithium Symposium, the industry was well ahead of the demand.

DEMAND AND FUTURE PROJECTIONS

The basic premise for the lithium industry and its capacity to provide enough lithium is the presumed demand for lithium batteries for electric cars. Today, it is impossible to reach a consensus on the market penetration of electric cars. Some have even predicted 130 million such cars by 2021, a difficult task since the projected global car production in 2020 is expected to reach 70.9 million.

According to J.D. Powers Associates, battery-powered and hybrid-electric cars will only capture around 7.3 percent of all car sales by 2020. That means that electric car sales will only account for 5.2 million car sales out of 70.9 million global car market. [19].

Figure 4 illustrates a similar prediction by TRU Group [20]

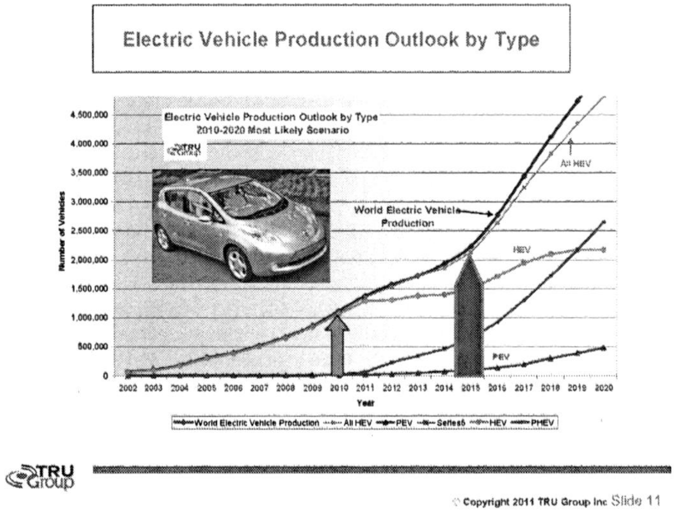

Figure 4 – Projection of Electric Vehicle Demand

Figure 5 summarizes the most likely demand scenario projected by the TRU Group [21], which predicted that the existing production capacities will exceed the demand by 2020.

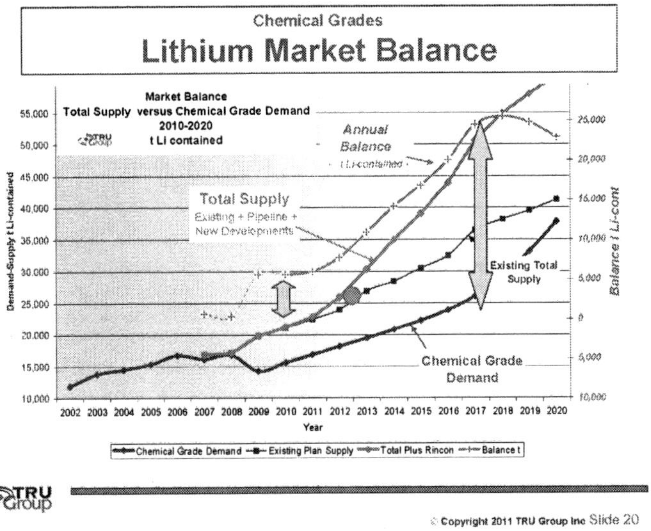

Figure 5 – Lithium Market Balance

A common misconception is that an electric car produces zero emission. An electric car requires regular recharging, which is provided by electricity from fossil fuels. A recent study on plug-in car mileage reported that a Nissan Leaf in Tucson AZ is the same as driving a gas-powered car that gets 48 miles to the gallon. In Colorado, it is apparently worse, being equivalent to a car, which gets 33 miles to the gallon. "We don't want to replace tailpipe emissions with the emissions of coal- and oil-fired power plants that produce the electricity used by BEVs," said Humphrey, Vice President of J.D. Powers Associates. Not a very good trade-off.

Furthermore, there has been some disconcerting recent news from car manufacturers. Toyota announced that it had abandoned to mass-produce electric vehicles (to the dismay of anti-oil activists). Toyota stated that it will release only 100!! eQ cars in the US and Japan. Two years ago, it had planned to release several thousands. Now only 21 hybrids will be released in 2015 [22]. Apparently, the company has a lower confidence in the battery technology, witness the recent bankruptcy filing by companies such as A123. [23]

CONCLUSION

Is lithium the metal of the future? As the author once asked: "QuoVadis, Litio"? , the existing lithium supplies will adequately satisfy the present and potential future lithium demand. The answer is now as it was then: "Io".

REFERENCES:

1. O'Neill, B. J., Kunasz, I.A., Wright, L.A., 1969, "The Lithium Production Curve: A Measure of Many Things", Earth and Mineral Sciences, Vol. 38, No. 6, The Pennsylvania State University, University Park, PA.
2. Talison Lithium Limited, 2011, N43-101 Technical Report;
3. Galaxy Resources Ltd., 2012, Ni43-101, Quarterly Report
4. Barrett, W.T., and O'Neill, B.J., Jr., 1970, "Recovery of Lithium from Saline Brines Using Solar Evaporation, " 3rd Salt Symposium, J.L. Rau and L.F. Dellwig, eds., Vol. 2, Northern Ohio Geological Society, Inc., Cleveland, pp. 47-50.
5. Moraga, B.A., et al, 1974, "Estudio Geologico del Salar de Atacama, Provincia de Antofagasta" Instituto de Investigaciones Geologicas de Chile, Bulletin 29, 56 pp.
6. Nicolli, H.B., et al., 1980, "Caracteristicas Geoquimicas de Aguas y Salmueras de la Puna Argentina," Miscelana No. 63, Academia Nacional de Ciencas, Cordoba, Argentina, pp. 3-38.
7. Ballivian, O., and Risacher, F., 1981, "Les Bassins a Evaporites de l'Altiplano Bolivien, ORSTOM, Paris, pp. 109-135.
8. http://www.westernlithium.com
9. http://www.simbolmaterials.com
10. http://www.albemarle.com
11. http://www.riotintominerals.com/.../RTM.JadarFactSheets.Sept2011.pdf
12. http://www.sqm.com - "U.S. Securities Exchange Commission Form 20F, 2011"
13. http://www.rockwoodspecialties.com
14. http://www.fmclithium.com
15. Kesler. T., 1978, Raw Lithium Supplies, Mining Engineering, Vol. XXX.
16. Evans, R.K., 2012, An Overabundance of Lithium? Industrial Minerals, Lithium Supply and Market Conference, Buenos Aires, Argentina
17. Kunasz, I. A., 2006, Lithium Resources, Industrial Minerals and Rocks: Commodities, Markets and Uses, Society for Mining, Metallurgy, and Exploration, pp. 601-613
18. Clarke, G., 2012, "Advancing lithium projects & expansions", Industrial Minerals, London.
19. http://venturebeat.com/2011/04/27/jd-power-electric-car-overhype/#m6eFU0exs5MBSMH2.99
20. Anderson, E R., 3rd Lithium Supply & Markets, 2011, "Shocking Future Battering the Lithium Industry through 2020", 3rd Lithium Supply & Market Conference, Toronto, Canada
21. Anderson E.R., 2012, "TRU Long-range supply-demand forecast for global lithium 2020". TRU Group, Toronto , Canada.
22. www.reuters.com, Sept 24 2012, "Toyota drops plan for widespread sales of electric car"
23. Petersen, J., 2012, "Electric Vehicle Battery Grants, 3 Years Of Disappointment And Failure" in http://www.seekingalpha.com.

Mater. Res. Soc. Symp. Proc. Vol. 1492 © 2013 Materials Research Society
DOI: 10.1557/opl.2013.615

Modified Carbon Papers as Electrode Materials of all-Vanadium Redox Flow Battery

Chih-Hsing Leu[1*], Shu-Yuan Chuang[1,2], Kan-Lin Hsueh[1], Jia-Ming Huang[1], Chia-Chun Chung[1], Chun-Hsing Wu[1], Hsiao-Hsuan Hsu[1], Yi-Ray Chen[1], Wen-Sheng Chang[1]

[1]*Green Energy and Environment Research Laboratories, Industrial Technology Research Institute, Hsinchu 310, Taiwan, Republic of China*

[2]*Department of Materials Engineer, Ming Chi University of Technology, Taipei 243, Taiwan, Republic of China*

* Corresponding author. Fax: +886-3-5820030, *E-mail address:* CHLeu@itri.org.tw

ABSTRACT

The electrode materials for VRFB should possess higher electric conductivity, corrosion resistance and hydrophilic properties in sulfuric acid. The characteristics of the electrode materials affect the stability and the energy efficiency of VRFB. Carbon materials are the best suited for VRFB applications. In this study, the calcined treatment, acid treatment and ozone treatment were used to modify the surface of carbon papers. The redox reaction of $[VO]^{2+}/[VO_2]^+$ on the modified carbon papers was evaluated by cyclic voltammetry (CV). The surface compositions of carbon materials were analyzed by X-ray photoelectron spectrometry (XPS). The experimental results reveal that three oxidative methods enhance the redox reaction of $[VO]^{2+}/[VO_2]^+$. The calcined treatments and acid treatments also enhanced hydrolysis reaction. The mole ratio of O/C apparently increased, but the binding energy of C1s and O1s were not chemically shifted in the acid treatment. The intensity of binding energy of O1s, between 532 eV and 534 eV, apparently increased in the ozone and calcined treatments. The Ox treated samples were more hydrophilic than the Oz treated samples. In the Ox treated samples, the decrease of R_{ct} value indicates that was contributed from the redox reaction of $[VO]^{2+}/[VO_2]^+$ and hydrolysis reaction. It does not completely benefit the energy efficiency of VRFB. The 5 x 5 cm^2 modified carbon papers were used as electrode materials in the VRFB. The voltage efficiency, coulomb efficiency and energy efficiency reached 93 %, 90 % and 83 %, respectively, at a current density of 12 mA . cm^{-2} at 0.8-1.8 V.

Keywords: Modified carbon paper ; Vanadium redox flow battery ; Energy storage ; Electrode materials ; Treatment

1. INTRODUCTION

The practical utilization ratio of renewable energy sources such as solar energy or wind energy can be promoted by finding compatible energy storage systems. All vanadium redox flow battery (VRFB) is one of potential energy storage technologies due to high energy efficiency, long cycle life and low cost etc. [1-4]. In general, carbon based materials [5-8] are used as the electrode materials of vanadium redox flow battery. Han et al. [9] revealed that carbon base materials were proved to show poor kinetic reversibility. Therefore, the surface modification of carbon electrode materials improves the performance of all vanadium redox flow battery. The –OH groups and carbonyl groups were introduced onto carbon materials by different oxidation methods. Sun et al. [2] proposed that $[VO]^{2+}/[VO_2]^+$ redox reaction would be strongly influenced by the concentration and nature of oxygen function groups on the electrode surface. Yue et al. [10] revealed that the activity of carbon fibres toward $[VO]^{2+}/[VO_2]^+$ redox reaction is improved by –OH groups. Li et al. [11] proposed that the increased activity for the vanadium redox reactions can be attributed to the increased surface and the concentration of function groups of –COOH.

This study used the calcined treatment, acid treatment and ozone treatment to modify the surface of carbon papers. The function groups were identified by X-ray photoelectron spectrometry (XPS). The redox reactions of $[VO]^{2+}/[VO_2]^+$ were measured by cyclic voltammetry (CV). The variation of binding energy of O1s and redox activity of $[VO]^{2+}/[VO_2]^+$ on the modified carbon papers were revealed. Finally, the performance of vanadium redox flow battery using the modified carbon papers as the electrode was done. Electrochemical impedance spectroscopy (EIS) of the modified carbon papers was monitored after charge-discharge cycles.

2. EXPERIMENTAL DETAILS

2.1 Treatments of carbon papers

The commercial 10AA carbon paper was as the raw material of the electrode of vanadium redox flow battery. First, the calcined treatment, the raw materials were calcined at 400 °C for 1 h, 2 h, 3 h, 4 h, 8h and 16 h, respectively. The treatment samples were noted as Ox xh. Second, the acid treatment, the raw materials were soaked in the 98 % H_2SO_4 for 1 h, 2 h, 4 h, 8 h and 12 h, respectively. The treatment samples were noted as Oh xh. Third, the ozone treatment, the raw materials were treated at 500-600 ppm ozone for 1 h, 2 h, 4 h, 6 h and 8 h, respectively. The treatment samples were noted as Oz xh.

2.2 XPS measurements

XPS spectra were obtained with ESCALAB250 XPS (Thermo Fisher Scientific, USA) at 2×10^{-9} mba. The maximum peak intensity of binding energy of C1s and O1s were normalized to

1000 and 10000, respectively.

2.3 Electrochemical measurements

The cyclic voltammograms were obtained using three-electrode electrochemical cell on Bisat (BioLogic Science Instruments). A thee-electrode cell was used with the carbon paper as the working electrode, a saturated Ag/AgCl electrode as the reference electrode and Pt wire as the counter electrode. The 9 mm diameter carbon paper fixed to touch 5 mm diameter glass carbon disk electrode used as the working electrode. The test solution was 0.2 M $VOSO_4$/ 2 M H_2SO_4 and the scan rate of a 20 mV/s ran from open circuit voltage (OCV).

The 5 x 5 cm^2 carbon papers were used as positive and negative electrodes. The separated membrane was Nafion 117 membrane (Dupont, USA). The initial catholyte and anolyte were 2 M V(IV) / 2 M H_2SO_4 and 2 M V(III)/ 2 M H_2SO_4. The volume of anolyte and catholyte were 10 and 12.5 ml. Electrochemical impedance spectroscopy (EIS) was measured on Autolab workstation (PGSTAT-30, Netherlands) over the frequency ranging from 10^5 to 0.1 Hz at the current of 300 mA. Constant current charge-discharge test was carried out by Autolab workstation (PGSTAT-30, Netherlands). The charge-discharge condition was set at constant current of 300 mA between the voltage of 1.8 V and 0.8 V.

3. RESULTS AND DISCUSSION

The commercial 10AA papers were used as the raw material of modified electrode due to the consistent property of raw carbon papers. The modified carbon papers were created to the different surface properties by three oxidation methods. This study explored the relation between the surface properties and the activity of the electrodes in the vanadium redox flow battery.

3.1 Redox reaction of $[VO]^{2+}/[VO_2]^+$

The modified carbon papers shows an electrocatalytic activity for $[VO]^{2+}/[VO_2]^+$ with the reversible redox potential at 0.60 V and 1.16 V. Figure 1 shows cyclic voltammetric behavior of the Ox 0 h, 1 h, 2 h, 3 h, 4 h, 8 h and 16 h samples. Following the increase of calcined treatment time, the electrocatalytic activity apparently increases during 1.2 V- 1.6 V. This is contributed to hydrolysis reaction. Table 1 lists the area ratio (vs. 10AA carbon paper) of oxidation, reduction and hydrolysis reaction. The integration area means the summation of faradic current between the reduction and oxidation currents. The difference between the area of oxidation and reduction reaction is attributed to hydrolysis reaction. The area of the electrocatalytic activity of $[VO]^{2+}/[VO_2]^+$ and hydrolysis reaction increases 1.8 and 19.2 times, respectively, after calcined treatment at 16 hours. Figure 2 shows cyclic voltammetric behavior of the Oh 0 h, 1 h, 2 h, 4 h, 8 h and 12 h samples. Following the increase of acid treatment time, the electrocatalytic activity of $[VO]^{2+}/[VO_2]^+$ slightly increases. The area of the electrocatalytic activity of $[VO]^{2+}/[VO_2]^+$ and hydrolysis reaction increases 0.22 and 2.1 times, respectively, after acid treatment at 12 hours (Table I). Figure 3 shows cyclic voltammetric behavior of the Oz 0 h

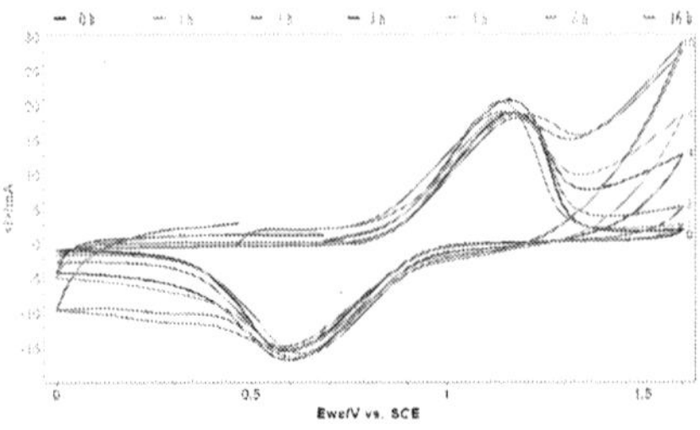

Figure 1. Cyclic voltammograms of the Ox 0h, 1h, 2h, 3h, 4h, 8h, 16h treated samples.

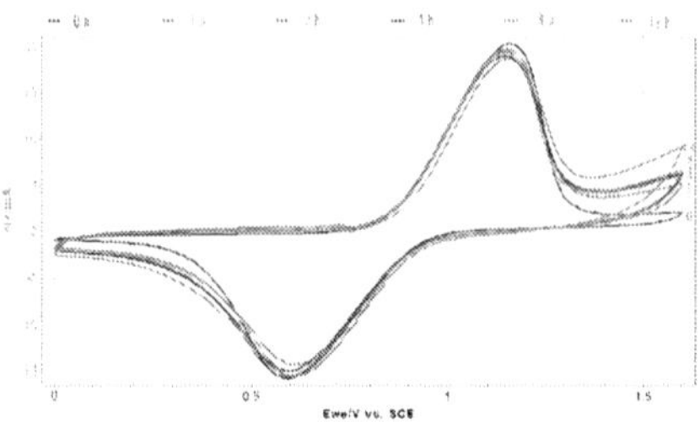

Figure 2. Cyclic voltammograms of the Oh 0h, 1h, 2h, 4h, 8h, 12h treated samples.

Table I Area ratio of oxidation and reduction of the modified carbon papers for CV tests.

Sample	1 (vs 10AA)*1	2 (vs 10AA)*2	1-2 (vs 10AA)*3
10AA	1.00	1.00	1.00
10AA-Ox 1h	1.04	1.02	1.34
10AA-Ox 2h	1.14	1.06	2.46
10AA-Ox 3h	1.67	1.38	6.63
10AA-Ox 4h	1.92	1.41	10.74
10AA-Ox 8h	2.69	1.84	17.51
10AA-Ox 16h	2.80	1.79	20.17
10AA-Oh 1h	1.04	1.00	1.81
10AA-Oh 2h	1.19	1.12	2.19
10AA-Oh 4h	1.17	1.10	2.28
10AA-Oh 8h	1.18	1.10	2.37
10AA-Oh 12h	1.34	1.22	3.09
10AA-Oz 1h	0.98	0.96	1.39
10AA-Oz 2h	1.05	1.03	1.36
10AA-Oz 4h	1.02	0.99	1.36
10AA-Oz 6h	1.18	1.17	1.36
10AA-Oz 8h	0.97	0.95	1.33

Note:
1. Oxidation area of sample/Oxidation area of 10AA
2. Reduction area of sample/Reduction area of 10AA
3. (Oxidation-Reduction) area of sample/(Oxidation-Reduction) area of 10AA

Table II The O/C ratio of the modified carbon papers.

Sample Name	O/C ratio
10 AA	0.02
10 AA-Ox 2h	0.02
10 AA-Ox 3h	0.03
10 AA-Ox 4h	0.03
10 AA-Ox 8h	0.02
10 AA-Oh 1h	0.09
10 AA-Oh 2h	0.10
10 AA-Oh 12h	0.13
10 AA-Oz 4h	0.03
10 AA-Oz 6h	0.02
10 AA-Oz 8h	0.02

, 1 h, 2 h, 4 h, 6 h and 8 h samples. The electrocatalytic activity of $[VO]^{2+}/[VO_2]^+$ increases 0.17 times at 6 hours and the ability of hydrolysis is almost no change (Table I).

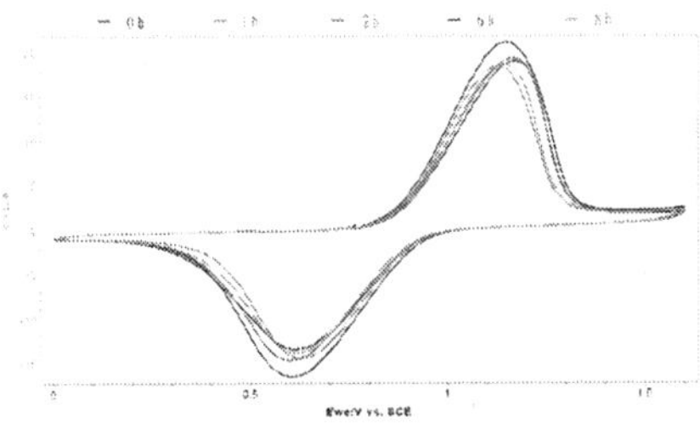

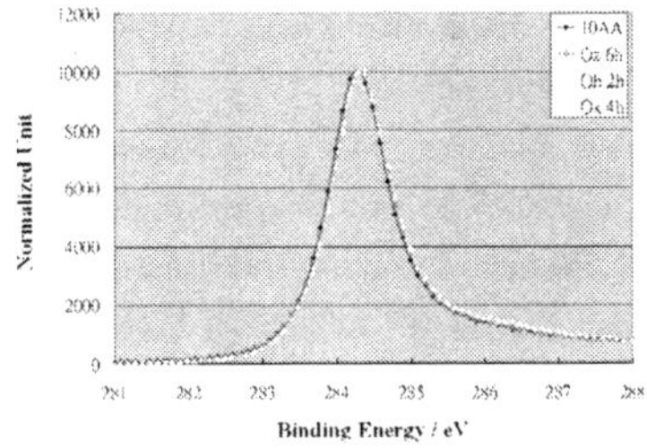

Figure 3. Cyclic voltammograms of the Oz 0h, 1h, 2h, 6h, 8h treated samples.

Figure 4. C1s binding energy of Oz 6h, Oh 2h, Ox 4h and 10AA samples.

3.2 Surface structure of the modified carbon papers

The mole ratio of O/C on the surface of the modified carbon papers under the calcined and ozone treatments is not apparent change but that under acid treatment apparently increases (Table II). It means that the calcined and ozone treatments were used to modify the type of function groups. The acid treatments were like to create the same type of function group similar to raw carbon paper. This study tries to use the normalization of maximum peak density instead of the simulated lines of peak separation in the literature. The peak intensity of C1s and O1s were normalized to compare chemically shifted of binding energy.

The binding energy of C1s is not change on the oxidation treatments (Figure 4). It can not reflect on the change of binding energy due to low carbon ratio of C-O function group. Figure 5 shows the binding energy of O1s during the calcined treatments. The peak becomes broad

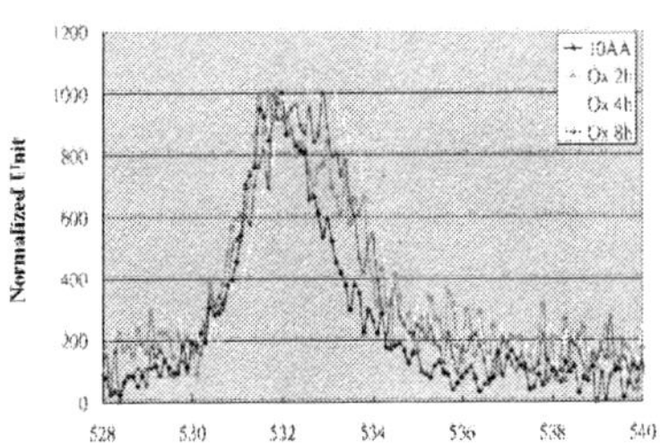

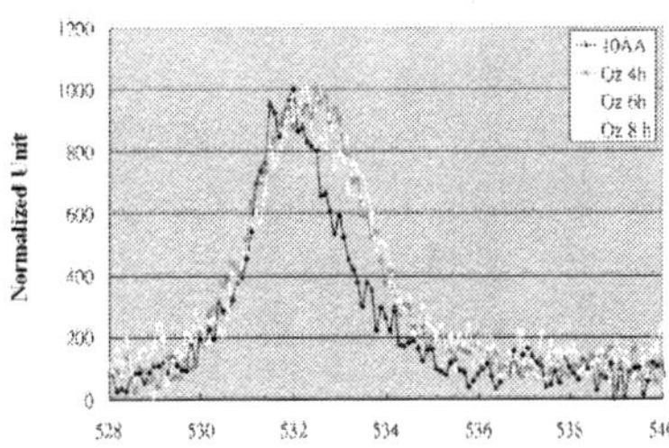

Figure 5. O1s binding energy of Ox 2h, Ox 4h, Ox 8h and 10AA samples.

Figure 6. O1s binding energy of Oz 4h, Oz 6h, Oz 8h and 10AA samples.

and the center of peak was chemically shifted from 532 ev to 533 eV. Figure 6 shows the binding energy of O1s on the ozone treatments. The center of peak was chemically shifted from 532 ev to 532.5 eV. There were not chemically shifted under the acid treatments (Figure 7). The literature reports [2, 10, 12, 13] revealed that peak 1 (about 531 eV), peak 2 (about 533 eV) and peak 3 (above 534 eV) are attributed to the O1s signals from C=O, C-OH and adsorbed water (or chemical absorbed oxygen), respectively. Overlapped the spectra of O1s binding energy of different oxidation treatments (Figure 8), the O1s peak of Ox treated samples is broader than that of Oz treatments between 532 and 534 eV. It means that the Ox treated samples were more hydrophilic than the Oz treated samples. The Oh treated samples were similar to raw 10AA carbon paper.

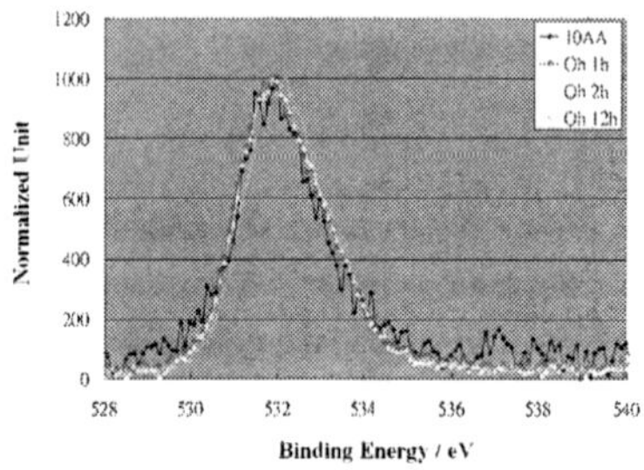

Figure 7. O1s binding energy of Oh 1h, Oh 2h, Oh 12h and 10AA samples.

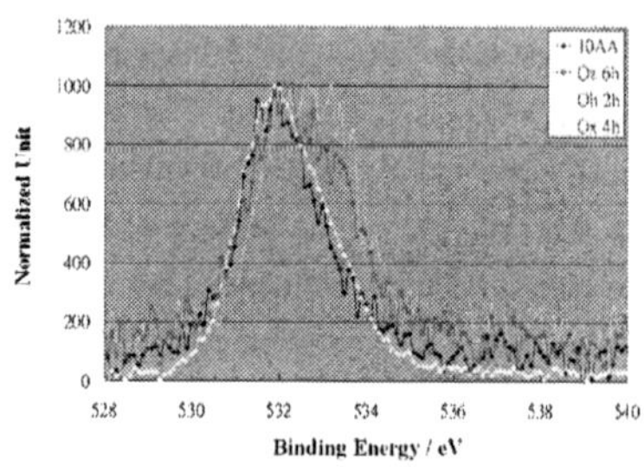

Figure 8. O1s binding energy of Oz 6h, Oh 2h, Ox 4h and 10AA samples.

3.3 Single cell battery test

Charge-discharge tests were performed by using a single cell with the modified carbon paper electrodes. Figure 9 shows the charge-discharge curve of VRFB using the Oz 6h treated

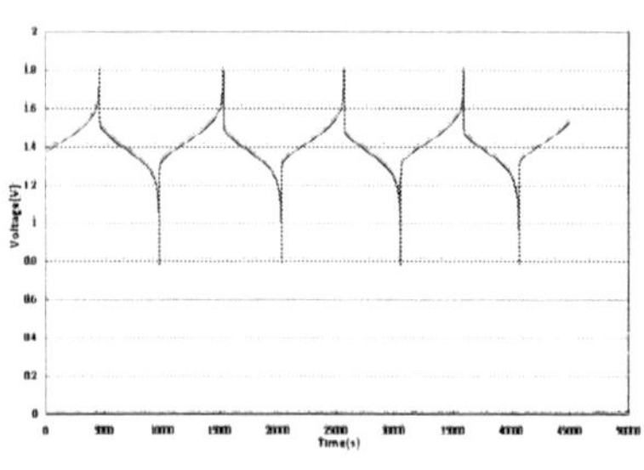

Figure 9. The charge-discharge curve of VRFB using the Oz 6h treated electrodes.

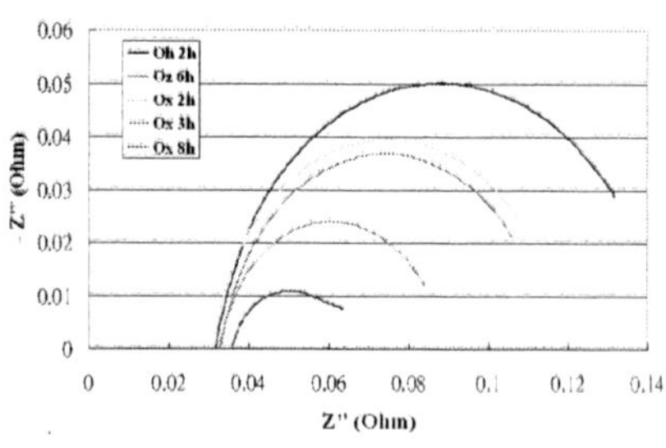

Figure 10. Nyquist plots of Oh 2h, Oz 6h, Ox 2h, Ox 3h and Ox 8h treated electrodes.

electrodes. All performed efficiency of the modified carbon paper electrodes lists in the Table 3.

Compared to raw 10AA carbon paper electrode, the performed efficiency of the Oh 2h treated sample was no apparent change. The Oz 6h, Ox 2h and Ox 3h treated electrodes can improve the energy efficiency of VRFB from 76 % up to above 80 %. Compared to the Ox treated electrodes, the energy efficiency of VRFB increased to 82 % using the Ox 3h treated electrodes and then that decreased to 79 % using the Ox 8h treated electrodes. Figure 10 shows that the EIS of the modified carbon paper electrodes after charge-discharge tests. Z' value (R_l, Table 3) at Z"=0 combines the solution electrodes and the contact resistance (ohmic resistance) [14, 15]. The R_l value of Ox 8h treated electrode slightly increases but that of other treated electrodes are almost no change. The first arc (high frequency range) in the Nyquist plots is due to the charge transfer reaction at the electrolyte/electrode interface [14, 15]. The diameter reflects the charge transfer resistance (R_{ct}, Table III). The R_{ct} value of the Oh 2h treated electrodes is no change

Table III The R_l value, R_{ct} value and efficiency of VRFB using the Oh 2h, Oz 6h, Ox 2h, Ox 3h and Ox 8h treated electrodes

Sample Name	R_l (mohm)	R_{ct} (mohm)	Energy Efficiency(%)	Coulomb Efficiency (%)	Voltage Efficiency (%)
10AA			76	88	87
10AA-Oh 2h	32	100	76	87	88
10AA-Oz 6h	33	73	83	90	93
10AA-Ox 2h	32	79	81	90	90
10AA-Ox 3h	33	48	82	91	90
10AA-Ox 8h	35	22	79	90	87

Note: 1. Electrodes : 5 x 5 cm^2 10AA modified carbon paper.

2. EIS Conditions: Current at 300 mA, Frequency from 100000 to 0.1 Hz.

3. Charge-discharge at constant current 300 mA. Electrolytes: 2 M V(III)/V(VI)/ 2 M H_2SO_4.

but that of the Oz and Ox treated electrodes decrease.

According to the said results, the hydrophilic surface structure benefited charge transfer reaction. From the results of CV and EIS, the redox reaction of $[VO]^{2+}/[VO_2]^+$ and hydrolysis reaction can be improved by the surface treatments. During charge-discharge processes, the redox reaction of $[VO]^{2+}/[VO_2]^+$ competed with hydrolysis reaction at higher voltage. The value of charge transfer resistance in the Ox 8h treated electrodes is half than that in the Ox 3h treated electrodes but the energy efficiency in the Ox 8h treated electrodes is lower than that in the Ox 3h treated electrodes. The Oz 6h treated electrodes had higher energy efficiency of VRFB due to no increase hydrolysis reaction after surface structure modification. The surface treatment does not completely benefit the energy efficiency of VRFB as hydrolysis reaction increases instead of the redox reaction of $[VO]^{2+}/[VO_2]^+$.

CONCLUSIONS

This study used the calcined treatments, acid treatments and ozone treatments to modify the

10AA carbon papers. The surface composition and the redox reaction of $[VO]^{2+}/[VO_2]^+$ were evaluated by XPS and CV measurements. The modified carbon papers were used as the electrodes of VRFB to measure the EIS and charge-discharge properties of VRFB. The results showed below:

1. The calcined treatments and acid treatments also enhanced hydrolysis reaction.
2. The order of pretreated effects on the redox reaction of $[VO]^{2+}/[VO_2]^+$ was the calcined treatments › the acid treatments › the ozone treatments.
3. The O/C ratio apparently increased, but the binding energy of C1s and O1s were not chemically shifted in the acid treatments.
4. The intensity of binding energy of O1s, between 532 eV and 534 eV, increased in the ozone and calcined treatments. The Ox treated samples were more hydrophilic than the Oz treated samples.
5. R_l value increases and R_{ct} value decreases following time increase of calcined treatments. The decrease of R_{ct} value indicates that was contributed from the redox reaction of $[VO]^{2+}/[VO_2]^+$ and hydrolysis reaction. It does not completely benefit the energy efficiency of VRFB.
6. The 5 x 5 cm^2 carbon papers were used as electrode materials in the VRFB. The voltage efficiency, coulomb efficiency and energy efficiency reached 93 %, 90 % and 83 %, respectively, at a current density of 12 mA . cm^{-2} at 0.8-1.8 V.

ACKNOWLEDGMENTS

This work was financially supported by a project of the Bureau of Energy, Ministry of Economic Affairs, Taiwan, Republic of China.

REFERENCES

[1] Kaneko H, Nozaki K,Wada Y, Aoki T, Negishi A, *Electrochem Acta* **36**, 1191(1991).

[2] B.T. Sun, M. Skyllas-Kazacos, *Electrochem. Acta* **37**, 2459(1992).

[3] M. Gattrell, J. Park, B. MacDougall, J. Apte, S. McCarthy, C. W. Wu, *J Electrochem Soc.* **151**, 123(2004).

[4] G. Oriji, Y. Katayama, T. Miura, *J Power Sources* **139**, 321(2005).

[5] M. Skyllas-Kazacos, *J. Power Sources* **124**, 299(2003).

[6] H.T. Zhou, H.M. Zhang, P. Zhao, B.L. Yi, *Electrochim. Acta* **51**, 6304(2006).

[7] Y.M. Zhang, Q.M. Huang, W.S. Li, H.Y. Peng, S.J. Hu, *J. Inorg. Mater.* **22**, 1051(2007).

[8] H.Q. Zhu, Y.M. Zhang, L. Yue, W.S. Li, G.L. Li, D. Shu, H.Y. Chen, *J. Power Sources* **184**, 637(2008).

[9] P. Han, H. Wang, Z. Liu, X. Chen, W. Ma, J. Yao, Y. Zhu, G. Cui, *Carbon* **49**, 693(2011).

[10] Lu Yue , Weishan Li , Fengqiang Sun , Lingzhi Zhao, Lidan Xing, *Carbon* **48**, 3079(2010).

[11] X. G. LI, K. L. Huang, S. Q. Liu, N. Tan, L. Q. Chen, *Trans. Nonferrous Met. Soc. China* **17**, 195(2007).

[12] S. Zhong, C. Padeste, M. Kazacos and M. Skyllas-Kazacos, *J. Power Sources* **45**, 29(1993).

[13] K. Jurewicz, K. Babe, *Energy Fuels*, **24**, 3429(2010).

[14] Y. Shao, M. Engelhard, Y. Lin, *Electrochemistry Communications* **11**, 2064(2009).

[15] Y. Shaoa, X. Wang, M. Engelhard, C. Wang, S. Dai, J. Liu, Z. Yang, Y. Lin, *J. Power Sources* **195**, 4375(2010).

Mater. Res. Soc. Symp. Proc. Vol. 1492 © 2013 Materials Research Society
DOI: 10.1557/opl.2013.471

Stability of Vanadium Electrolytes in the Vanadium Redox Flow Battery

Shu-Yuan Chuang[1,2], Chih-Hsing Leu[1*], Kan-Lin Hsueh[1], Chun-Hsing Wu[1], Hsiao-Hsuan Hsu[1], Yi-Ray Chen[1], Wen-Sheng Chang[1]

[1]*Green Energy and Environment Research Laboratories, Industrial Technology Research Institute, Hsinchu 310, Taiwan, Republic of China*

[2] *Department of Materials Engineer, Ming Chi University of Technology, Taipei 243, Taiwan, Republic of China*

* Corresponding author. Fax: +886-3-5820030, *E-mail address:* CHLeu@itri.org.tw

ABSTRACT

The stability of the negative electrode electrolyte affects the efficiency and capacity of energy storage in the vanadium redox flow battery (VRFB) system. To explore the stability of vanadium electrolytes, the study prepared five types of V(II) electrolytes that were exposed to air in a fixed open area and monitored the charge state of vanadium ions over time by UV/Visible spectrophotometer. This study succeeded in preparing pure V(II) electrolytes. Five characteristics are found in the UV/Visible spectra, respectively, during the oxidation process from V(II) electrolytes to V(III) electrolytes and V(III) electrolytes to V(IV) electrolytes. The experimental results show that the oxidation rate of a solution of 1 M V(II) electrolytes to V(III) electrolytes and 1 M V(III) electrolytes to V(IV) electrolytes under an atmosphere of air is 4.79 and 0.0089 mol/h per square meter. The oxidation rates of 0.05-1 M V(II) electrolytes to V(III) electrolytes are approximately 96-538 times than that of V(III) electrolytes to V(IV) electrolytes.

***Keywords*:** Electrolyte ; UV/Visible spectra ; Energy storage ; Vanadium redox flow battery

1. INTRODUCTION

The concept of redox flow cell was proposed by Thaller et al. [1] in 1974. Skyllas-Kazacos et al. [2-4] proposed the all vanadium redox flow battery (VRFB) in 1985. Compared to varied electric storage technologies, the VRFB is one of most promising storage energy systems considering the capital cost, the power rating and discharge time [5, 6]. In Japan and South Africa, the VRFB system closed to commercialization in load leveling, wind power and solar power [7] according to an expected long cycle life for the rechargeable cell [8, 9].

Compared to alternative battery solutions, the VRFB has unique and highly competitive features including no cross contamination, easy scale-up the battery storage capacity at low capital cost, high energy efficiencies possible, long cycle life, minimal safety etc. [10]. However, a number of side reactions can occur during operation and can cause loss of capacity over extended charge–discharge cycling including the oxidation of vanadium ion in air, gassing side reaction during charging, differential transfer of vanadium ion or volumetric transfer of electrolytes from one half-cell to the other [10]. It is easy to find that the violet V(II) oxidizes to the green V(III) in a short time in air. During charge-discharge cycles, the energy efficiency and battery capacity of VRFB will lose due to the oxidation of vanadium electrolytes in air.

The UV/Visible spectrophotometer is a good tool to monitor the concentration of vanadium ion but the UV/Visible spectra of vanadium electrolytes are inconsistent in the literatures [11, 12]. First, this study is to prepare different concentration V(II) electrolytes and monitor the oxidation processes of V(II) electrolytes. The standard UV/Visible spectra of V(II) and V(III) electrolytes were obtained during the oxidation processes of V(II) electrolytes. Second, the oxidation rate of V(II) and V(III) electrolytes were calculated by the decrease of V(II) concentration and the increase of V(IV) concentration, respectively. The variation of UV/Visible spectra of V(II) electrolytes oxidizing to V(IV) electrolytes and the ratio of oxidation rate of V(II) electrolytes to V(III) electrolytes were revealed.

2. EXPERIMENTAL DETAILS

2.1 V(II) Electrolytes Preparations

0.05 M, 0.1 M, 0.2 M, 0.5 M and 1 M $VOSO_4$/ 2 M H_2SO_4 were prepared as the starting electrolytes. First, the initial anolyte and catholyte were $VOSO_4$/ 2 M H_2SO_4 electrolytes in the VRFB system. It was charged at a current density of 10 mA . cm^{-2} and cut off at 1.8 V. The V(III)/ 2 M H_2SO_4 electrolytes were generated by electro-reduction of V(IV)/ 2 M H_2SO_4 electrolytes. Second, the V(IV)/ 2 M H_2SO_4 electrolytes were instead of the V(V)/ 2 M H_2SO_4 electrolytes (the catholyte). It was charged again at a current density of 10 mA . cm^{-2} and cut off at 1.8 V. The V(II)/ 2 M H_2SO_4 electrolytes were generated by electro-reduction of V(III)/ 2 M H_2SO_4 electrolytes. Then, the V(II)/ 2 M H_2SO_4 electrolytes were sealed in the UV cell.

2.2 UV/Vis Measurements

The UV/Visible spectra of the V(II)-V(IV)/ 2 M H_2SO_4 electrolytes were measured using a Evolution 201 UV/Visible spectrophotometer (double beam, Thermo) with a scan rate of 1200 nm/min from 1000 nm to 200 nm. A 2 M H_2SO_4 solution was used as a reference. First, the V(II)/ 2 M H_2SO_4 electrolytes sealed in the UV cell were measured by UV/Visible spectrophotometer. Then, the V(II)/ 2 M H_2SO_4 electrolytes in the UV cell were opened in air and the UV/Visible spectrophotometer was used to monitor the oxidation state of vanadium electrolytes following exposed time in air.

3. RESULTS AND DISCUSSION

3.1 V(II) electrolytes oxidizing to V(III) electrolytes

Figure 1 shows the UV/Visible spectra of 0.2 M V(II) electrolyte oxidizing to V(III) electrolyte. It is found that five characteristics in the varied UV/Visible spectra of the oxidation of V(II) electrolyte following the increase of oxidation time.

1. The trough at 350 nm shifts to 330 nm.
2. Three intersected points occur at 488 nm, 544 nm and 720 nm.
3. The peak at 368 nm shifts to 400 nm and the absorbance of the peak increases.
4. The peak at 568 nm shifts to 608 nm and the absorbance of the peak increases.
5. The peak at 848 nm does not shift and the absorbance of the peak at 848 nm decreases. Finally, it disappears.

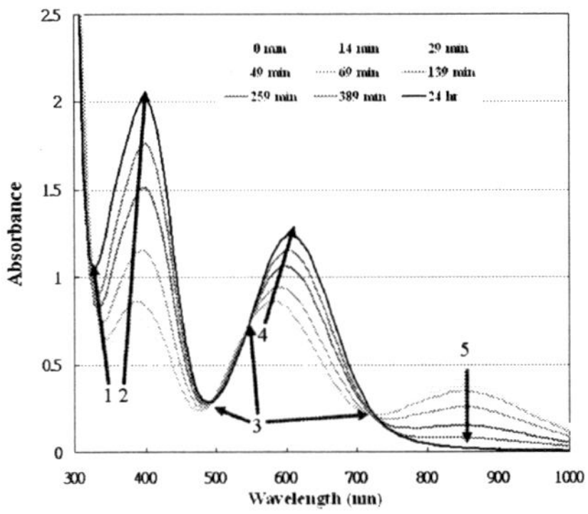

Figure 1. The UV/Visible spectra of V(II) electrolyte oxidizing to V(III) electrolyte.

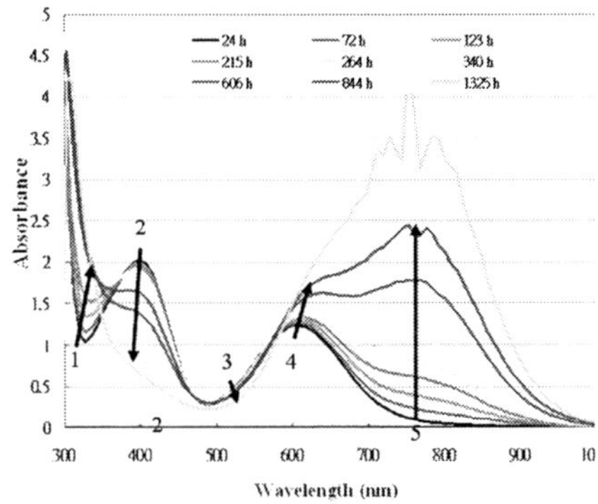

Figure 2. The UV/Visible spectra of V(III) electrolyte oxidizing to V(IV) electrolyte.

3.2 V(III) electrolytes oxidizing to V(IV) electrolytes

Figure 2 shows the UV/Visible spectra of 0.2 M V(III) electrolyte oxidizing to V(IV) electrolyte. Five characteristics are found in the varied UV/Visible spectra of the oxidation of V(III) electrolyte following the increase of oxidation time.

1. The trough at 330 nm shifts to 350 nm and then the trough disappears.
2. The absorbance of the peak at 400 nm decreases and then the peak disappears.
3. The absorbance between 490 and 580 nm slowly decreases.
4. The peak at 610 nm shifts to 640 nm. The absorbance of the peak increases.
5. The absorbance of the peak at 760 nm grows.

3.3 Calibration curves of V(II) and V(IV) electrolytes

Figure 3 displays that the UV/Visible spectra of 0.05 M, 0.1 M, 0.2 M, 0.5 M and 1 M V(II) electrolytes. Three peaks were found at 368 nm, 568 nm and 848 nm. The R^2 value of curve fitting of the peak absorbance at 368 nm, 568 nm and 848 nm versus V(II) concentration are

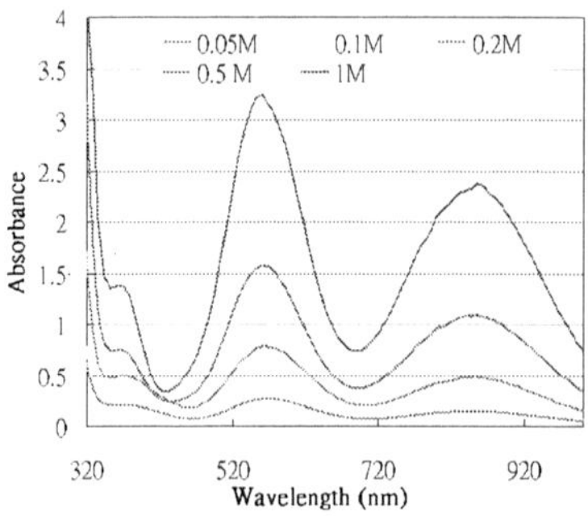

Figure 3. The UV/Visible spectra of 0.05 M- 1 M V(II) electrolytes.

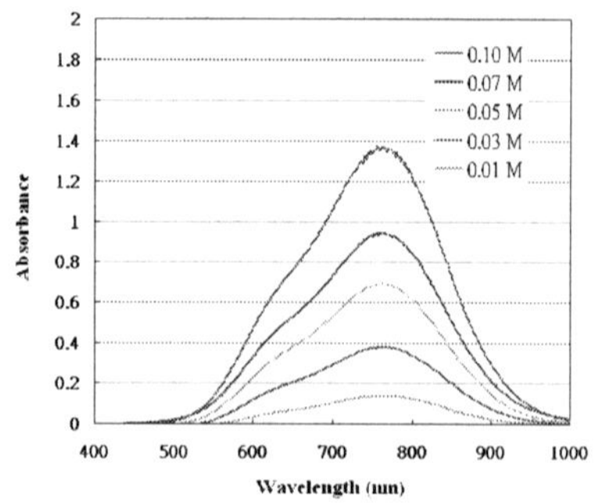

Figure 4. The UV/Visible spectra of 0.01 M- 0.1 M V(IV) electrolytes.

0.9880, 0.9979 and 0.9981, respectively. It is good correlation between V(II) concentration versus the peak absorbance at 568 nm and 848 nm. Considering the mixing electrolytes of V(II) and V(III), the peak absorbance at 848 nm is good selection for the concentration measurements of V(II) electrolytes due to no peak shift and no overlap peak absorbance. The relationship between the peak absorbance at 848 nm and the concentration of V(II) electrolyte is

$$V(II)\ C_M=(Abs_{848}+0.0001)/2.3035 \quad (1)$$

The charge state of V(III) electrolytes is between the V(II) and V(IV) electrolytes. Considering the stability of V(III) electrolytes, it is difficult to prepare the specific concentration of V(III) electrolytes. The variation of V(III) concentration in air can be monitored by the variation of V(IV) concentration. Figure 4 displays the UV/Visible spectra of 0.01–0.1 M $VOSO_4$ / 2 M H_2SO_4 electrolytes. Two peaks were found at 640 nm and 760 nm. The R^2 value of curve fitting of the peak absorbance at 640 nm and 760 nm versus V(IV) concentration are 0.999 and 0.999, respectively. It is good correlation between V(II) concentration versus the peak absorbance at 640 nm and 760 nm. Considering the mixing electrolytes of V(III) and V(IV), the peak absorbance at 760 nm is good selection for the measurements of V(IV) electrolytes due to no peak shift and no overlap peak absorbance. The relationship between the peak absorbance at 760 nm and the low concentration of V(IV) electrolytes is

$$V(IV)\ C_M=(Abs_{760}+0.0062)/13.606 \quad (2)$$

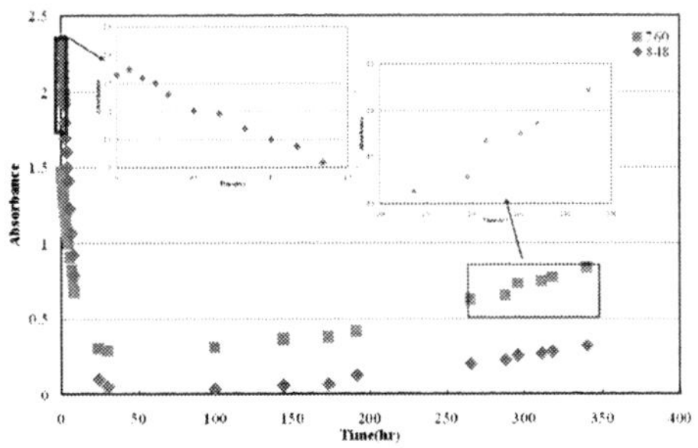

Figure 5. The absorbance at 760 nm and 848 nm versus oxidation time.

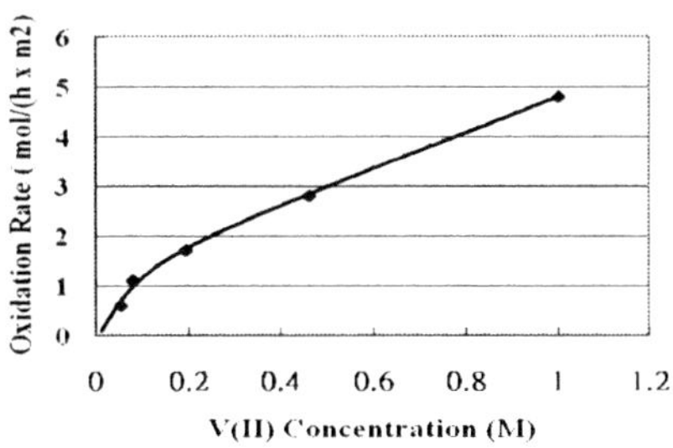

Figure 6. The relationship between the oxidation rate and the concentration of V(II) electrolytes.

3.4 Oxidation rate of V(II) and V(III) electrolytes

The absorbance of the peak at 848 nm versus the oxidation time of 1 M V(II) electrolyte is shown in Figure 5. The absorbance is linearly decrease after exposed to air for initial 60 minutes. The oxidation rate of 1 M V(II) electrolyte is calculated by the slope of the V(II) concentration versus the oxidation time according to equation (1). The oxidation rate of 1 M V(II) electrolyte is 4.79 mol/(h·m^2). The oxidation rate of other V(II) concentrations can be calculated by this method. The oxidation rate of V(II) electrolytes in air versus the V(II) concentration is shown in Figure 6. The oxidation rate of V(II) electrolytes in air is linearly increase following the V(II) concentration over 0.2 M.

The characteristic peaks of V(III) electrolyte at 400 nm and 608 nm are mixing with that of V(IV) electrolyte during the oxidation process. The variation of V(III) concentration can be calculated by the formation of V(IV) concentration at a absorbance of 760 nm. The absorbance at 760 nm versus the oxidation time of 1 M V(II) electrolyte is shown in Figure 5. The maximum slope of the V(IV) formation concentration versus the oxidation time is used to calculate the oxidation rate of approximate 1 M V(III) electrolyte according to equation (2). The oxidation rate of approximate 1 M V(III) electrolyte is 0.0089 mol/(h·m^2). The oxidation rate of other V(III) concentrations can be calculated by this method. The oxidation rate of V(III) electrolytes in air versus the V(III) concentration is shown in Figure 7. The oxidation rate of V(III) electrolytes in air is between 0.0056 mol/(h·m^2) and 0.011 mol/(h·m^2). No specific trend was found between the oxidation rate and the concentration of V(III) electrolytes.

The ratio of the oxidation rate of V(II) electrolytes to V(III) electrolytes is shown in Figure 8. The oxidation rates of 0.05 M, 0.1 M, 0.2 M, 0.5 M and 1 M V(II) electrolytes to V(III) electrolytes are approximately 96, 155, 158, 503 and 538 times than that of V(III) electrolytes to V(IV) electrolytes, respectively. The higher concentrations of V(III) electrolytes are more stable than V(II) electrolytes in air.

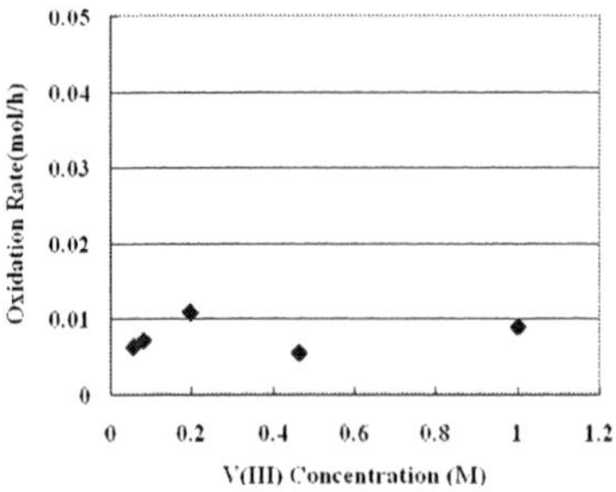

Figure 7. The relationship between the oxidation rate and the concentration of V(III) electrolytes.

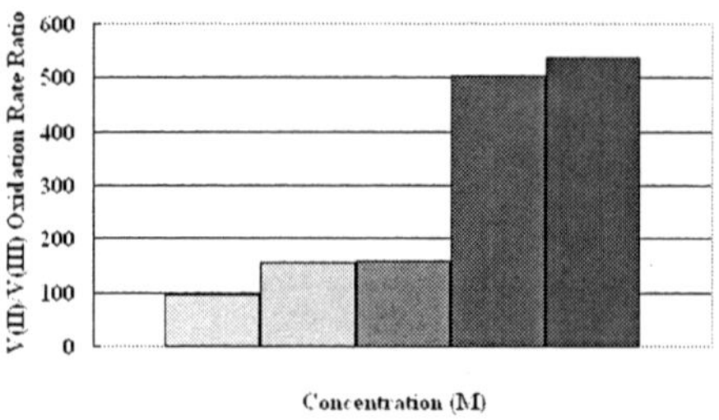

Figure 8. The ratio of oxidation rate of 0.05 M- 1 M V(II) electrolytes to V(III) electrolytes

Compared to the spectrum of V(II) electrolyte in the literature [12] , the peak absorbance at 848 nm was not found. The peak absorbance at 760 nm was found in the V(III) electrolyte. From the results of Fig. 1 and 2, the study infers that the V(II) and V(III) electrolytes were not pure for UV/Visible measurements in the literature [12]. The V(II) and V(III) electrolytes were partially oxidized before UV/Visible measurements. The V(II) electrolyte showed the spectrum of mixed V(II) and V(III) electrolytes. The V(III) electrolyte showed the spectrum of mixed V(III) and V(IV) electrolytes. The prepared electrolytes must be isolated from air before UV/Visible measurements.

CONCLUSIONS

This study successes to monitor the oxidation process of V(II) electrolytes by UV/Vis spectrophotometer in air. During the oxidation process, the standard UV/Vis spectra of V(II) and V(III) electrolytes were acquired. Other information revealed below:

1. Five characteristics are found in the UV/Vis spectra, respectively, during the oxidation process from V(II) electrolytes to V(III) electrolytes and V(III) electrolytes to V(IV) electrolytes in air.
2. The concentration of V(II) and V(VI) ions can be monitored by the UV/Vis absorbance at 848 nm and 760 nm.
3. The oxidation rates of 0.05 M-1 M V(II) electrolytes to V(III) electrolytes 96-538 times than that of V(III) electrolytes to V(IV) electrolytes. The higher concentrations of V(III) electrolytes are more stable than V(II) electrolytes in air.

This study suggests that the negative electrode electrolytes were protected by the inert gas to prevent the reduction in energy efficiency and storage capacity of VRFB.

ACKNOWLEDGMENTS

This work was financially supported by a project of the Bureau of Energy, Ministry of Economic Affairs, Taiwan, Republic of China.

REFERENCES

[1] L. H. Thaller, NASA TMX-71540, National Aeronautics and Space Administration: Washington, DC, 1974.

[2] E. Sum, M. Rychcik, M. Skyllas-kazacos, *J Power Sources* **16**, 85(1985).

[3] E. Sum, M. Rychcik, *J Power Sources* **15**, 179(1985).

[4] M. Skyllas-Kazacos, M. Rychcik, R. G. Robins, A. G. Fane, *J Electrochem Soc* **133**, 1057 (1986).

[5] Electrical Storage Association. http://www.electricitystorage.org/ESA/technologies/.

[6] H. Chen, T. N. Cong, W. Yang, C. Tan, Y. Li, Y. Ding, *Prog. Natural Sci.* **19**, 291(2009).

[7] P. Han, H. Wang, Z. Liu, X. Chen, W. Ma, J. Yao, Y. Zhu, G. Cui, *Carbon* **49**, 693 (2011).

[8] M. Gattrell, J. Park, B. MacDougall, J. Apte, S. McCarthy, C. W. Wu, *J Electrochem Soc.* **151**,123(2004).

[9] C. Fabjan, J. Garche, B. Harrer, L. Jorissen, C. Kolbeck, F. Philippi, *Electrochem Acta* **47**, 825(2001).

[10] M. Skyllas-Kazacos, M. Kazacos, *J. Power Sources* **196**, 8822(2011).

[11] X. W. Wu,"High performance and stabilization electrolytes for vanadium redox flow battery", Central South University Master Thesis(2010).

[12] M. Vijayakumar, M.S. Bhuvaneswari, P. Nachimuthu, B. Schwenzer, S. Kim, Z. Yang, J. Liu, L. G. L. Graff, S. Thevuthasan, J. Hu, *Journal of Membrane Science* **366**, 325(2011).

Rare Earths and Magnets

Mater. Res. Soc. Symp. Proc. Vol. 1492 © 2013 Materials Research Society
DOI: 10.1557/opl.2013.174

The Elements of Magnetics

Steve Constantinides
Arnold Magnetic Technologies Corporation, 770 Linden Avenue, Rochester, NY 14625, U.S.A.

ABSTRACT

The 20th century saw rapid and dramatic improvements in permanent magnet materials. It has been 31 years since the discovery of neodymium-iron-boron and numerous companies and laboratories are seeking to produce a new and superior material. Topics discussed herein are material options, economics of selected materials and market drivers in material selection. Market issues include manufacturability by shape, size, and material yield; raw material supply including cost and dependability of the supply chain; raw material and magnet product price stability; development of applications based on commercial needs, government legislation and consumer demand. "Need is the mother of invention" and no discussion would be complete without covering why a new material would be beneficial from an applications point of view especially in energy production and consumption. Therefore, an introduction will be provided for select, major applications using permanent magnets and the growth forecasts for these.

INTRODUCTION

Research in magnetic materials has been conducted for a few hundred years. Permanent magnets have especially benefitted from advances in key magnetic properties. The 20th century saw particularly significant improvements. [1] The last substantially improved material has been so superior to any that came before that it has been widely adopted and not surpassed. This material is neodymium-iron-boron (NdFeB, "Neo"). It was discovered in 1981 by accident at the Naval Research Laboratories during research for an improved soft magnetic material. Three patents were issued for it including process, composition and product patents. The material was then optimized and commercialized by both Sumitomo and General Motors and numerous other companies in Europe, the USA and China. The first commercial magnets were sold by Crucible Magnetics of Elizabethtown, Kentucky, in November of 1984.

The first major application for Neo was in hard disk drives. By 1990, it was estimated that 75% of Neo production was for the sintered magnets used to move the read/write head and for the compression bonded ring magnet in the spindle motor of hard disk drives. Other motors also benefitted from Neo magnets' combination of high magnetic output and relatively low price.

During the 1980s and early 1990s almost all the rare earths used in magnets came from Molycorp's Mtn. Pass California mine. The mine was operated primarily for other rare earths such as europium for color cathode ray tubes and cerium for glass polishing. The magnet rare earths were sold from the residual materials after separation. By the mid 1990's the demand for neodymium exceeded what was extracted in natural ratio causing stockpiling of cerium and lanthanum.[2] By the late 1990s China had become a major supplier of rare earths (oxides, metals and alloys) eroding Molycorp's domination of the market.[3]

By the great recession of 2009, almost 97% of rare earths were coming from China and rising prices and threats of inadequate supply caused market panic. Thus developed a great interest in rare earths and in alternatives to rare earths, especially in magnets.

DISCUSSION

The word "elements", as presented here, refers first to the fundamental science of magnetism. Understanding the principles of magnetism is necessary for understanding the market and its responses to supply disruption. The second "element" refers to those market forces that drive utilization of one material versus another and to what extent a material is used at all. We'll look at three markets: vehicular traction drives, wind power generation of electricity, and electronics – especially memory storage devices. The next element refers to those factors controlling material selection – the economics of materials. A side issue, but one that has been very important with regard to Neo magnets during 2011 is the threat of non-supply, a total cut-off of the magnets or raw materials used to make Neo to companies and consumers outside China. Then we come to the elements themselves. Which ones have been used in magnetic materials, which elements might be used and what alloys are being investigated which is the last element of the discussion – research and development - efforts to find the next permanent (or soft) magnet.

Elemental Magnetism

A good place to begin when speaking about magnetism is with the magnetic hysteresis loop. In response to an externally applied field, magnetic materials generate an internal field – an induced field. As we apply a stronger field in one magnetic direction (+H), weaken it and then apply it in the opposite direction (-H) we cause the internal field to respond accordingly. If we plot the applied field (H) versus the induced field (B), we have what is called the hysteresis loop of the magnetic material.[4,5]

We measure the combination of applied and induced field and represent it as the "Normal Hysteresis Loop". When we subtract the magnitude of the applied field from the combined field, the result is the "Intrinsic Hysteresis Loop". As we cycle the magnet around the hysteresis loop (by changing the applied field), we see that the 1st and 3rd quadrant are identical, though of reverse sign just as the 2nd and 4th quadrants are identical but of reverse sign.

The magnetic hysteresis loop provides a great deal of information about the material. First, it identifies the type of magnet: soft or permanent. A magnetically "soft" material has very small H_{cB} and the area circumscribed by the hysteresis loop is small. On the other hand, a permanent magnet exhibits a much greater H_{cB} and a larger area within the loop. Values of B_s represent the magnetic saturation and H_s the applied field strength at which saturation occurs.

The energy to force a magnet around the loop is proportional to the area within the loop. Therefore, lamination steel in transformers and motors benefits from being made of magnetically soft materials with very low H_{cB} and correspondingly small areas within the loop.

When Ferrite permanent magnets were invented in the mid 1950's, suddenly we had a material with H_{cJ} much larger than H_{cB} and it became common practice to provide both the Normal and the Intrinsic curves in product literature. This continued, of course, with SmCo, Neo and SmFeN magnets.

A list of the key characteristics of commercial permanent magnets includes:

- Flux density (Br)
- (Maximum) Energy Product (BHmax)
- Resistance to demagnetization (HcJ)
- Usable temperature range
- Magnetization change with temperature (RTC)
- Demagnetization (2nd quadrant) curve shape
- Recoil permeability
- Corrosion resistance
- Physical strength
- Electrical resistivity
- Magnetizing field requirement
- Available sizes, shapes, and manufacturability
- Material availability and product cost

For each application a subset of these characteristics will determine how well a material is suited to the application. The most discussed figure of merit is "maximum energy product", abbreviated BHmax (or "$(BH)_{max}$"). The amount of energy we can temporarily store in a magnetic circuit is proportional to a magnet's maximum energy product. Since these are supposed to be "permanent" magnets, we would not want them to demagnetize, therefore the value of H_{cJ}, which is a measure of resistance to demagnetization, is also important.

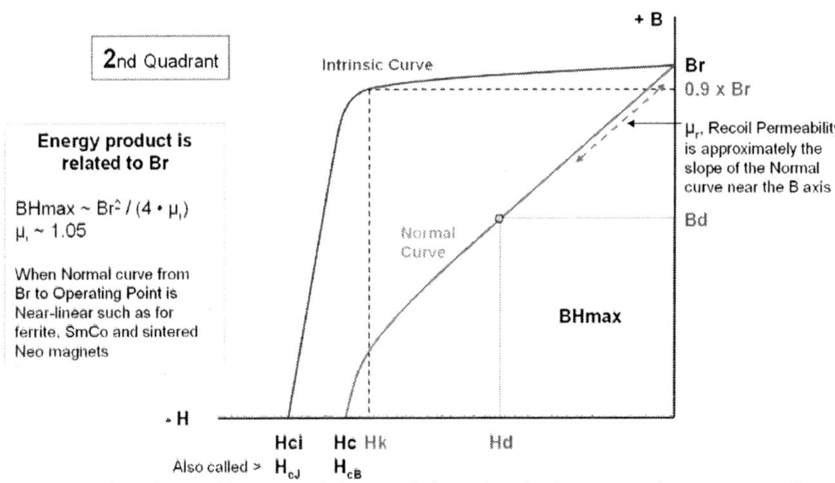

Figure 1. Second quadrant of the magnetic hysteresis loop showing key magnetic parameters of permanent magnets.

The following four commercially available magnet materials represent over 95% of permanent magnet sales: 1) Alnico, 2) Ferrite, 3) SmCo (1:5 and 2:17) and 4) Neo. Each has advantages that make it more suitable for a specific application. The values shown in Table I represent a practical range of available properties for each material. A satisfactory new material will have a propitious combination of these values as indicated in the rightmost column.

Table I. Values for the key characteristics of commercially available permanent magnets. The right column provides indicators of what would be a better magnet material.

Characteristic	Units	Alnico 5-7	Alnico 8	Ferrite 8	Ferrite 9	SmCo 1:5	SmCo 2:17	NdFeB 33EH	NdFeB 48M	Indicator
Flux density (Br)	Tesla	1.35	0.85	0.39	0.45	0.9	1.1	1.15	1.39	> is better
Energy Product (BHmax)	kJ/m^3	60	46	28	37	175	230	230	370	> is better
Resistance to demagnetization (Hcj)	kA/m	59	125	245	370	2400	1600	2400	1115	> is better
Usable temperature range	°C	4 K to 520 °C	4 K to 520 °C	-40 to 150 °C	-40 to 150 °C	4 K to 520 °C	4 K to 520 °C	150 K to 200 °C	150 K to 100 °C	minimum -40 to 200 °C
Induction change with temperature (RTCof Br)	%/°C	-0.02	-0.01	-0.2	-0.18	-0.045	-0.035	-0.11	-0.12	< is better
2nd quadrant Normal curve shape		Curved	Curved	Straight	Straight	Straight	Straight	Straight	Straight	Straight
Recoil permeability	B / H	2	2	1.04	1.04	1.03	1.05	1.04	1.05	~1
Corrosion resistance		Excellent	Excellent	Outstanding	Outstanding	Good	Good	Fair	Fair	Outstanding
Physical strength	MPa	55	205	65	70	120	120	285	285	> 50 also "tough"
Electrical resistivity	eπ • cm	47	50	10^6	10^6	55	90	180	180	> is better
Magnetizing field requirement	kA/m	120	240	480	800	2000	4000	2700	2700	Less than 4000
Coefficient of Termal Expansion	%/ °Cx10^{-6}	11.5	11	10 to 15	10 to 15	7 to 14	11 to 13	7.5 to -0.1	7.5 to -0.1	< 15
Approx Current Price (ballpark estimates)	$/kg	$40	$35	$8	$15	$120	$100	$200	$150	< is better
Relative Cost at 20 °C	$/MGOe	$5.3	$6.1	$2.3	$3.2	$5.5	$3.5	$6.9	$3.2	< is better
Relative Cost at 200 °C	$/MGOe	$5.7	$6.3	$5.6	$7.1	$6.5	$3.9	$10.8	n/a	< is better

Elements driving growth of the magnet market

Since 2010, major topics of interest within the magnet industry have been pricing and availability of rare earths and rare earth magnets. The family of Rare Earth magnets includes $SmCo_5$, Sm_2Co_{17}, SmFeN and NdFeB ("Neo").

SmCo has been discussed as a replacement for Neo magnets. Although SmCo magnets are superior for elevated temperature applications, the raw materials used in Neo magnets are more generally available and have been historically at lower cost. This has propelled Neo magnets into a dominant position. However, for Neo to perform successfully at elevated temperature requires substituting dysprosium for up to 1/3 of the neodymium (up to ~11 weight percent of a magnet). In years 2011 and 2012, the supply of dysprosium has become inadequate, resulting in extraordinarily high material prices. There is likely to be a continuing long-term shortage as no significant new supplies of dysprosium have been identified.

SmFeN decomposes at a fairly low temperature, starting at ~450 °C, preventing consolidation to full density. Because it must be used as a (low temperature formable) bonded magnet, maximum energy product is limited by dilution with a non-magnetic binder. Therefore it is a niche product.

Table II is a break down of applications for rare earth magnets with approximate percentages going into each category. It represents about 80% of the market for rare earth magnets. Much of the missing information is with regard to the domestic markets in China, India and other countries where definitive market information has not been available.

Table II. Applications for rare earth magnets with approximate fraction of all applications and estimates of requirements for neodymium and for dysprosium.[6]

	2010				2015			
	yr 2010	**Magnet**	Oxide, tons		yr 2015	**Magnet**	Oxide, tons	
Applications	**% of mix**	tons	**Nd**	**Dy**	**% of mix**	tons	**Nd**	**Dy**
Motors, industrial, general auto, etc	25.5%	15,871	7,122	1,059	25.0%	24,316	10,912	1,622
HDD, CD, DVD	13.1%	**8,140**	4,196	0	14.4%	**14,040**	7,237	0
Electric Bicycles	9.1%	**5,680**	2,549	379	8.2%	**7,955**	3,570	531
Transducers, Loudspeakers	8.5%	5,290	2,727	0	6.5%	6,322	3,259	0
Unidentified and All Other	6.5%	4,046	1,995	90	6.0%	5,836	2,878	130
Magnetic Separation	5.0%	3,112	1,466	138	3.4%	3,307	1,558	147
MRI	4.0%	2,490	1,228	55	1.5%	1,459	720	32
Torque-coupled drives	4.0%	2,490	1,117	166	2.5%	2,432	1,091	162
Sensors	3.2%	1,992	982	44	1.5%	1,459	720	32
Hysteresis Clutch	3.0%	1,867	879	83	1.5%	1,459	687	65
Generators	3.0%	1,867	769	194	1.0%	973	400	101
Energy Storage Systems	2.4%	1,494	670	100	2.5%	2,432	1,091	162
Wind Power Generators	2.1%	**1,300**	583	87	10.1%	**9,810**	4,402	654
Air conditioning compressors and fans	2.0%	1,245	559	83	2.5%	2,432	1,091	162
Hybrid & Electric Traction Drive	0.9%	**570**	214	80	6.3%	**6,160**	2,308	867
Misc: gauges, brakes, relays & switches, pipe inspection, levitated transportation, reprographics, refrigeration, etc.	7.7%	4,792	2,186	285	7.1%	6,906	3,113	447
Total	100.0%	**62,246**	29,243	2,843	100.0%	**97,296**	45,037	5,115

Nd: 54% increase

Dy: 80% increase

REO requirement includes 80% oxide to metal, 97% metal alloying, and 80% magnet manufacturing material yields.

There are significant geographic differences in product mix depending on where the downstream products are made and what market segment is being served. China now manufactures about 80% of all rare earth magnets, Japan makes 17%, and about 3% are made in Europe and the USA.

Because of raw material supply price variability and questions about availability, numerous magnet-using companies have changed product plans: product roll-outs have been delayed, new designs are being re-evaluated to use alternate magnet materials and technologies without permanent magnets are being investigated and utilized.

The change in magnet usage between 2010 and 2015 suggests a requirement for an increase in supply of neodymium (oxide and metal) of about 54% while dysprosium requirements will need to increase 80%. Without adequate dysprosium, neo magnet growth will be suppressed. To understand the market dynamics better, let's look at a few specific applications – those in bold face in Table II.

Cars and small trucks are increasingly being driven by some combination of electric motor (traction drive motor) with or without an internal combustion engine (ICE). Forecast sales in the USA of cars and small trucks using alternate drive systems for 2012, based on sales through August, are 461,250 units. 2011 and 2012 represent watershed years in that EV (electric vehicle) and PHEV (plug-in hybrid electric vehicle) sales have just started. Change in consumer preferences are due in large part to the high price of gasoline, overall improvement in economic prosperity, market acceptance of alternate drive systems and the increasing number of models to appeal to a wider consumer audience.

HEV/EV/PHEV sales, as a percentage of overall sales in the USA, are increasing from 2.25% of sales in 2011 to 3.14% in 2012. 54 vehicle models are being marketed. There should be no doubt that we are at a major turning point regarding use of hybrid and full electric vehicles.

A second developing market for permanent magnets is wind power – both generation of commercial electricity by large, tower-mounted generators and smaller systems for home and business installations. In 2008, DOE forecast installations[7] of commercial electric wind power generation. Actual installations have exceeded forecast largely due to government incentives. Future installations will depend upon continuing incentives in combination with higher alternate fuel costs.

Countries outside the USA have also been installing wind power with China leading the surge, installing 43% of new wind power in 2011. Prior to 2005, China's wind power industry was close to non-existent. Since then it has grown rapidly and now cumulatively China has more wind power generation installed than any other country (26.2% of global generation).

Top 10 cumulative capacity Dec 2011

Country	MW	% SHARE
PR China	62,364	26.2
USA	46,919	19.7
Germany	29,060	12.2
Spain	21,674	9.1
India	16,084	6.8
France**	6,800	2.9
Italy	6,737	2.8
UK	6,540	2.7
Canada	5,265	2.2
Portugal	4,083	1.7
Rest of the world	32,143	13.5
Total TOP 10	205,526	86.5
World Total	**237,669**	**100.0**

** Provisional Figure

Source: GWEC

Top 10 new installed capacity Jan-Dec 2011

Country	MW	% SHARE
PR China	17,631	43
USA	6,810	17
India	3,019	7
Germany	2,086	5
UK	1,293	3.2
Canada	1,267	3.1
Spain	1,050	2.6
Italy	950	2.3
France**	830	2.0
Sweden	763	1.9
Rest of the world	4,865	12.0
Total TOP 10	35,699	88
World Total	**40,564**	**100.0**

** Provisional Figure

Source: GWEC

Figure 2. Global wind power installations[8]

Consensus estimates of the fraction which are permanent magnet (PM) type generators recently installed are 5% in the US and Europe and 25% in China. These numbers are highly speculative as producers are reluctant to make public information about generator type. Reasonable estimates for kilograms of Neo required per megawatt are 600 kg/MW in direct drive and 200 kg/MW in hybrid drive generators.

One of the main uses for Neo rare earth magnets is in hard disk drives (HDD), CDs and DVDs where a bonded Neo magnet is used for driving the spindle motor and sintered magnets are used for positioning the read/write head and providing a clamping force (in some CDs and DVDs). Even though the amount used per drive is small, 6 to 13 grams, the huge quantity of devices requires large amounts of rare earth magnets. Importantly, these devices use no dysprosium. In Figure 3 we see continuing growth of the HDD market despite the surge in SSD (solid state drives, thumb drives).

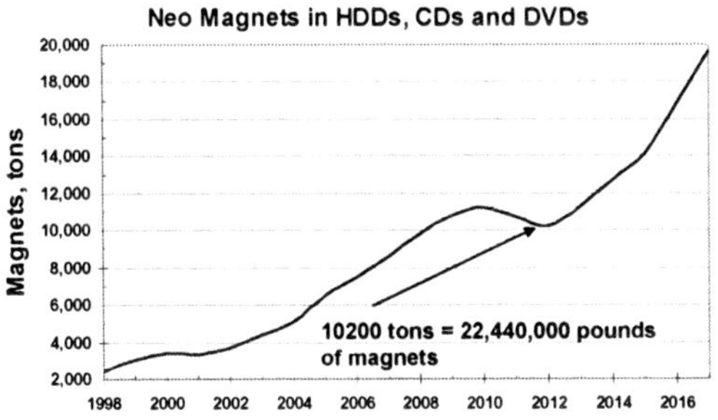

Figure 3. Consumption of Neo magnets in HDDs, CDs and DVDs[9]

Economic Elements

An examination of product selling price of each of the four major commercial permanent magnets highlights the high cost of rare earth magnets – primarily due to the cost of the raw materials. Many factors such as shape, complexity, size and country of origin contribute to a magnet's manufacturing and selling price. The values shown in Figure 4 are fair estimates for simple shapes and intermediate sizes. Selling price is that observed in the USA and Europe and provides comparative values.

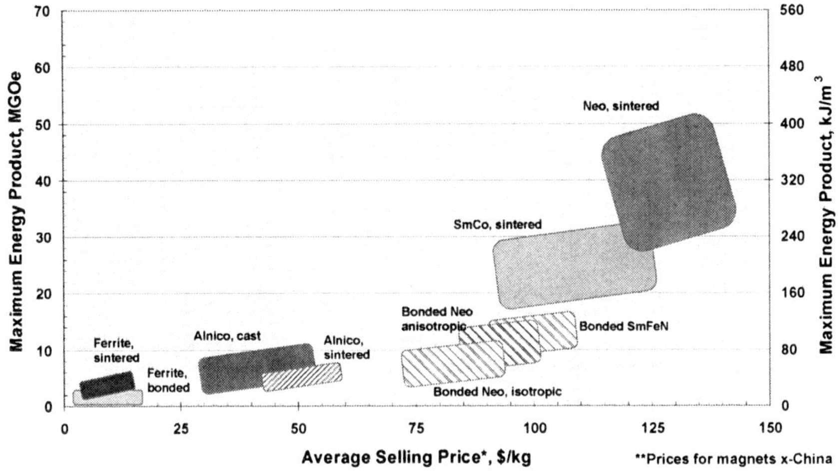

Figure 4. Relative pricing of the most common commercially available permanent magnets. Values represent pricing as of October 2012.

Rare earth materials have experienced price inflation and market disruption.(10) When prices became too high, users of rare earth magnets switched to alternate designs and are now slow to return. Reduction in demand is helping to speed a drop in material prices.

While prices of rare earths are coming back down, they will probably not reach levels seen at the low point in 2003. For example, a slow continuing increase is expected in SmCo magnet prices. A stabilization of Neo magnet pricing at higher prices than SmCo is expected due primarily to a combination of high demand and continued high pricing of dysprosium. Of course, prices can always change due to government policies, raw material discoveries, new magnetic material invention, changes in market needs, consumer sentiment, etc.

The Elements

Any discussion of commercial viability has to start with the premise that the raw materials are readily available and at a reasonable cost. As a primary ingredient, it's highly recommended to select more common materials such as those above the upper dashed line in Figure 5. Minor ingredients may be from between the upper and lower dashed lines. But elements from below the lower dashed line should be avoided except in the very smallest additions.

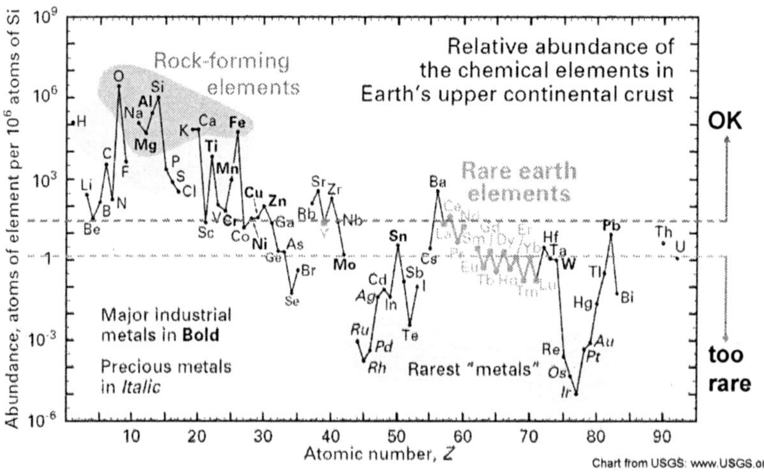

Chart from USGS: www.USGS.org

Figure 5. Relative availability of the elements referenced to silicon, the most abundant element(11)

Let's work with the periodic table to see what elements are likely candidates for use in magnetic materials. I will use a method similar to that of Dr. William McCallum of Ames Laboratory who kindly shared his notes with me several months ago. (I take personal responsibility for material presented here).

Let's start thinning the list of elements to remove synthetic (man-made) elements, radioactive elements, noble gases – the inert elements, toxic materials, and let's get rid of the elements that are truly rare such as platinum, palladium, gold, silver, etc. Let's also remove the rock-forming elements and hydrogen. So we're down from 90 naturally occurring elements to 36 – still a large number to research – shown in Figure 6.

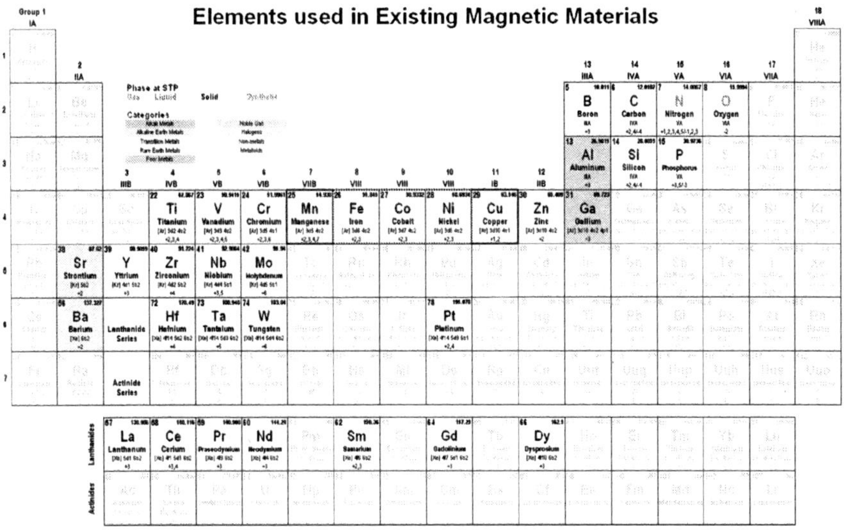

Figure 6. Periodic Table of the Elements with only those elements highlighted that have been most suitable for making magnetic materials. This conforms to the residual elements after removing those that are not suitable for magnetics with the exception of Pt, Ge and Sn.

Table III. Magnet materials and the common constituents of those materials

	Major constituents						Minor constituents				Comments
Soft Magnetic Materials											
Iron	Fe										Low carbon mild steel
Silicon Steel	Fe						Si				Si at 2.5 to 6%
Nickel-Iron	Fe	Ni									Ni at 35 to 85%
Moly Permalloy	Ni	Fe					Mo				Ni at 79%, Mo at 4%, bal. Fe
Iron-Cobalt	Fe	Co					V				23 to 52% Co
Soft Ferrite	Fe	Mn	Ni	Zn			O				
Metallic Glasses	Fe	Co	Ni				B	Si	P		Amorphous and nanocrystalline
Permanent Magnets											
Co-Steels	Fe	Co									
Alnico	Fe	Ni	Co	Al	Cu		Ti	Si			
Platinum Cobalt	Pt	Co									
Hard Ferrites	Fe	Sr									Oxygen dilutes; Ba no longer used
SmCo	Co	Sm	(Gd)	Fe	Cu	Zr					
Neodymium-iron-boron	Fe	Nd	Dy	(Y)	B	Co	Cu	Ga	Al	Nb	
Cerium-iron-boron	Fe	Nd	Ce	B							Limited use in bonded magnets
SmFeN	Fe	Sm	N								Nitrogen is interstitial; stability issue
MnBi	Mn	Bi									Never commercialized
MnAl(C)	Mn	Al					C				Not successfully commercialized

What elements have been used over the last 150 years to make magnetic materials? The list in Table III contains most common magnetic materials and the elements contained in them. If we create a periodic table with these elements highlighted, it is the same as the distilled table above (Figure 5) except for these exceptions: 1) Platinum: platinum-cobalt was the first high performance magnet. It was used to make watch drive motor magnets whose very small size compensated for the high material cost. It is still made but in limited quantities. 2) Germanium and Tin have not been used, at least to my knowledge, in commercial magnets except as very minor additives. Like aluminum and gallium, they might make suitable modifying constituents to assist sintering or phase formation.

Elements in Research

I've split existing research projects into two categories. The first could be called "Variation on a Theme" as it represents an extension of research on materials that have been previously examined. A listing of the materials being researched includes:

- SmCo plus exchange-coupled soft phase
- NdFeB plus exchange-coupled soft phase
- Fe-N (variation of SmFeN, interstitial N)
- Mn alloys: MnBi, MnAl(C)
- Heusler alloys
- Alnico – modified to enhance coercivity
- Carbides: FeC, CoC, Co,TM-C
- Modified Ferrites (chemical or structural modifications): La-Co Ferrites, Core-Shell structure ferrites
- Ce-Co,Fe and Ce-Fe,Co-B,C

Although some excellent research and development took place on these materials over extended periods several differences exist between what took place "then" and what is being pursued "now". One difference is that current analytic capabilities are superior to what existed even two or three decades ago either as a new technology or greatly improved one. Secondly, we now have techniques to form these materials with a refined structure at micro- and nano-scales, building larger structures a few atoms at a time. And we have methods of simultaneously creating numerous alloy specimens through co-sputtering techniques greatly speeding evaluation of promising alloy systems.

The second grouping of projects is what I call the "bottoms-up" approach. It is to take the 36 remaining elements from our experiment with the periodic table and to combine them using computer algorithms to forecast the potential for generating a useful magnetic moment. From the total list of candidates, a short list will be parsed and these most promising alloys produced and evaluated.

A significant hurdle is to make a nano-particulate material fully dense and to do so in a safe, scalable, and economic manner. A number of compaction/consolidation techniques exist: Hydraulic or mechanical compaction in a die, CIP (cold isostatic pressing), Explosive compaction, Liquid phase sintering, Warm/Hot compaction (liquid phase enhanced), Hot forging, HIPing (hot isostatic pressing), Rolling (with tension), Hot extrusion with induced stress/strain, SPS (Spark Plasma Sintering), and Friction stir welding. It will be necessary to determine the most successful route to full density.

The beneficial properties of magnetic materials are due in part to either (crystal) shape anisotropy (e.g. alnico) or magneto-crystalline anisotropy (e.g. ferrite, SmCo and Neo). In either case, during manufacture, the magnetic domains must be mutually aligned to obtain maximum properties. Nano-particulate rare earth magnet alloys have existed for over a decade. However, simultaneous densification and alignment has been a problem awaiting solution.

In addition to the importance of structure in current research is the importance of thermal processing in the development of optimal microstructure and magnetic properties. With the exception of ceramic (hard ferrite) magnets, magnetic alloys are just that – alloys. Therefore, thermal treatments to form a stable and desirable phase structure are common in existing products. In Figure 7, Karl Strnat shows the development of the hysteresis loop of Sm_2Co_{17} during its thermal treatment. Sm_2Co_{17} is a complex alloy also containing Fe, Cu and Zr. The final heat treatment is required to develop a stable structure which is a combination of the 2:17 phase and a grain boundary phase of SmCo 1:5.

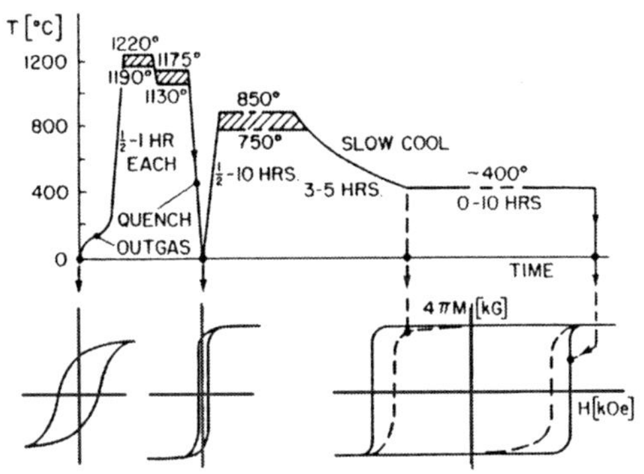

Figure 7. Typical temperature profile for the sintering and heat treating of "2-17"-type $Sm(Co,Fe,Cu,Zr)_{7.2-8.5}$ magnets [16]

We might say that the right composition provides the opportunity for good magnetics and the right thermal treatment creates the right alloy phase structure and optimum magnetic properties.

SUMMARY

Looking once more at the price chart for magnetic materials, Figure 4, research is focused on either improving the properties of less expensive materials or reducing the cost of the high performing materials, all while using readily available raw materials. These new magnetic materials will be necessary not as a replacement for but in addition to existing magnetic products. Almost all previously manufactured permanent magnet materials are still in production. New material will likely impact but not eliminate existing materials.

Breadth of acceptance of a new material will depend on many factors not the least of which is manufacturing cost. Ferrite magnets still represent about 85% of all permanent magnets made (on a weight basis) even though they are only 10% of the energy product of Neo magnets. The reasons: low cost and wide material availability.

New materials may result from new elemental combinations. But it is more likely that structure and thermal processing will play the major role in new magnetic materials.

ACKNOWLEDGMENTS

William McCallum of Ames Laboratory who first presented the periodic table element review method and who graciously shared that with me.

REFERENCES

1. R.J. Parker, Advances in Permanent Magnetism, (John Wiley & Sons, Inc., 1990) p.331
2. J. Morrow (Private communication)
3. G.B. Haxel, J.B. Hedrick, G.J. Orris, *Rare Earth elements – Critical Resources for High Technology*, (USGS Fact Sheet 087-02, 2002) p.1
4. J. Wernick (Bell Telephone Laboratories), *Kirk-Othmer Encyclopedia of Technology, vol 15, Bulk Materials* (John Wiley & Sons, Inc., 1995) p.45
5. ASTM A977-07 – *Standard Test Method for magnetic properties of high coercivity permanent magnet materials* (ASTM, West Conshohockin, PA, 2007) p.3
6. Numerous industry sources; W.T. Benecki, T.C. Clagett, S.R. Trout, *The Global Permanent Magnet industry* (self-published, 2010)
7. *20% Wind Energy by 2030: Increasing Wind Energy's Contribution to U.S, Electricity Supply, Executive Summary* (U.S. Department of Energy, December 2008) p.7
8. *The Global Wind Report Annual Market Update 2011* (Global Wind Energy Council, 2011) p.12
9. Data from iSuppli, Disk/Trend and Peripheral Research Corporation
10. S. Constantinides, *The Erratic but Continuing Demand for Rare Earth Materials in Permanent Magnets* (Vehicle Technologies Program 2013 Kick-off Meeting, Oak Ridge national Laboratories, 13 November 2012)
11. G.B. Haxel, J.B. Hedrick, G.J. Orris, *Rare Earth elements – Critical Resources for High Technology*, (USGS Fact Sheet 087-02, 2002) p.3
12. Periodic Table of the Elements from Vertex http://www.vertex42.com/ExcelTemplates/periodic-table-of-elements.html
13. R.M. Bozorth, *Ferromagnetism* (IEEE Press, 1993) p.328
14. J.M.D. Coey, *Magnetism and Magnetic Materials* (Cambridge University Press, 2009) p.394
15. P. J. Webster and K. R. A. Ziebeck, *Heusler Alloys*, in Landolt-Bernstein New Series Group III, Vol. 19C, H. R. J. Wijn (Ed.) (Springer, Berlin, 1988) p. 75
16. K.J. Strnat, *Rare Earth-Cobalt Permanent Magnets*, (Ferromagnetic Materials, vol 4, E.P. Wohlfarth and K.H.J. Buschow, ed., Elsevier Science Publishers B.V., 1988) p.184
17. C.D. Graham, Jr., *Magnetic Properties of Metals and Alloys* (American Society for Metals, 1958) p.307
18. Stanek et al, *Investigations of Thermo-Magnetic Treatment of Alnico 8 Alloy* (Archives of Metallurgy and Materials, Vol 55, 2010 Issue 2) p.573

Mater. Res. Soc. Symp. Proc. Vol. 1492 © 2013 Materials Research Society
DOI: 10.1557/opl.2013.543

Photothermal microscopy applied to the characterization of UO_2-Gd_2O_3 nuclear fuel pellets

Oscar Martínez[2,3,5] , Facundo Zaldivar[1], Nélida Mingolo[1], Rodolfo Kempf[4]

[1] Fisica, Facultad de Ingeniería, Universidad de Buenos Aires, Buenos Aires, CABA, Argentina.
[2]Fisica, FCEN, Universidad de Buenos Aires, Buenos Aires, CABA, Argentina.
[3] Toket SRL, Buenos Aires, CABA, Argentina.
[4] Unidad Actividad Combustibles Nucleares. División Caracterización, CNEA, San Martin, Buenos Aires,
Argentina.
[5] CONICET, Buenos Aires, CABA, Argentina

ABSTRACT

The photothermal photodeflection technique is shown to provide information on the homogeneity of fuel pellets, pore distribution, clustering detection of pure urania and gadolinea and to provide a two-dimensional mapping of the thermal diffusivity correlated to the composition of the interdiffused Gadolinium and Uranium oxide. Histograms of the thermal diffusivity distribution become a reliable quantitative way of quantifying the degree of homogeneity and the width of the histogram can be used as a direct measure of the homogeneity. These quantitative measures of the homogeneity of the samples at microscopic levels provides a protocol that can be used as a reliable specification and quality control method for nuclear fuels, substituting with a single test a battery of expensive, time consuming and operator dependent techniques.

INTRODUCTION

The development of manufacturing processes compatible with production requirements and quality restrictions gives rise to the need for detection and characterization techniques that are reliable, fast and inexpensive capable of determining as many relevant parameters as possible in a single test. In particular the UO_2-Gd_2O_3 to be used in new small frame power plants like the one under design by the CNEA (Atomic Energy Agency from Argentina) have stringent requirements regarding the homogeneity of the resulting solid solution, the size distribution of clusters of remaining pure oxides and pore sizes and distributions [1,2].

In this work a technique is presented that satisfies all these requirements, substituting with advantages in cost, time, accuracy and simplicity an entire battery of techniques otherwise necessary and currently in use such as X-ray diffraction and scanning electron microscopy employed to analyze the chemical homogeneity in UO_2-Gd_2O_3 pellets [3] and micrographic techniques performed to produce an image of the pellet surface after the sample is cut, polished and color etched, so that in the image the free gadolinia, urania and solid solution grains reflect different colors [3].

EXPERIMENT

Samples

Two different batches of (U, Gd)O_2 pellets were produced and the samples obtained were named Type I and Type II. The UO_2+Gd_2O_3 powders were blended by mixing the precursors in double cone blender, with an intensifier bar, rotating at 32 rpm, for 4 h. The UO_2 used to prepare all samples was an ex-ammonium uranyl carbonate powder with impurity content less than 150 ppm, specific surface area (BET) of 5.4±0.1 m^2/g, average particle size of 8.6±0.2 μm (Sedigraph), and O/U ratio of 2.05±0.01. The Gd_2O_3 raw powder was of commercial grade (i.e., 99,9 % pure), with average particle size of 2.7±0.2 μm, picnometry density 7g/cm^3, specific surface area (BET) of 2.7±0.1 m^2/g and was calcined at 900 °C during 2 hours before mixing.

The pellets were pressed in composition batches, where "green" pellets were obtained with pressure 300 MPa at ambient temperature. The pressing process was done in a floating table press, having green densities in the range 51-53 % theoretical density.

The two samples prepared differed in the sintering cycle. The sintering cycle was conducted under a 99.999 % H_2 atmosphere, at a flow of 200 ml/min, in molybdenum furnace, raising the temperature up to 1650 °C during 2 hours for type I and 1750 °C for 8 hours for type II. For the photothermal characterization the samples were encapsulated and polished to obtain an optically flat and reflective surface.

The technique

The technique used is based on a method reported before [4-6] consisting of heating the surface by a modulated laser beam and the simultaneous detection of the reflectance change [7-10], due to the temperature rise, and the deflection of the probe beam [11-12], due to the surface deformation caused by the localized thermal expansion. The deflection is sensed by intercepting part of the beam with a knife edge, generating in this manner a modulation of the detected signal as the beam oscillates laterally.

For these samples the contribution of the photodeflection is much larger than that of the change in reflectivity, only the dominant mechanism needs to be taken into account and the detected signal (modulated component of the reflected probe at frequency ω) can be used to retrieve the thermal diffusivity from a single measurement at a fixed modulation frequency as was shown in a previous work [6].

The procedure is to scan the surface at constant modulation frequency and build three images. One is the total reflected probe power which corresponds to the optical image of the sample. The other two images are the amplitude and phase delay of the photodeflection signal. From these three images a new set is generated with the following procedure. Pores are recognized as dark regions in a binarized optical image with a suitable selection of the threshold. After subtracting such regions form the amplitude image, a new threshold is established for which the dark regions correspond to pure Gadolinea clusters. This is because the Gadolinea has an optical bandgap in the UV, and does not absorb the pump laser radiation, hence yielding no thermal expansion signal [13]. For the remaining region (neither pore nor gadolinea) the thermal diffusivity is retrieved from the phase of the photothermal signal-[6]. The new set of images is presented in Figure 1 for two samples denoted from the two fabrication batches (Type I and Type II).

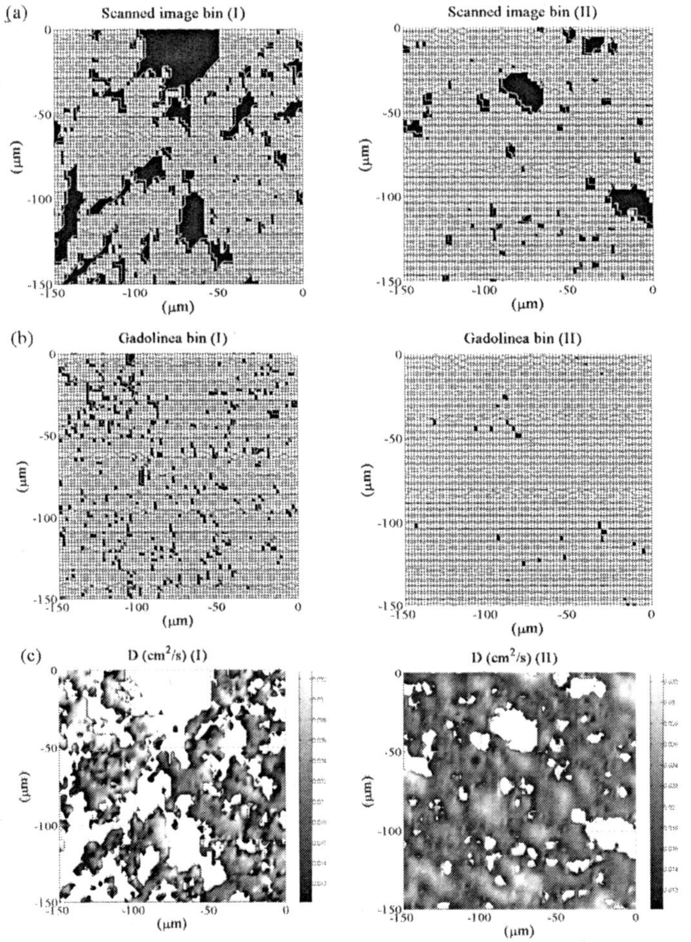

Figure 1. Processed images for sample I (left) and II (right). a) Binarized image for the pore distribution (dark). b) Binarized image for the Gadolinea clusters (dark). c) Thermal diffusivity map as retrieved from the phase of the photothermal signal.

The use of the photothermal signal to distinguish the homogeneity of the sintered material relies in the following two facts: one is that urania and gadolinea have very different thermal diffusivities, with $D(UO_2)= 0.027 cm^2/s$ [14] and $D(Gd_2O_3)=0.118\ cm^2/s$, (measured by the flash method in a sample prepared for this purpose) and hence a mixture of the two materials should have a thermal diffusivity somewhere in between, i.e. larger than that of urania. The second one is that the interdiffused sample has a thermal diffusivity lower than that of urania as determined for samples with 10% Gd by the flash method [15], yielding $D(UO_2+10\%\ Gd_2O_3)=0.0156\ cm^2/s$ which is more than 40% lower than urania, allowing the clear distinction between the mixture and the interdiffusion on one side and a determination of the homogeneity, as the method used has a resolution of about 5% (understanding by resolution the distinction within one scan, while accuracy is of about 15% and is a comparison with a standard).

DISCUSSION

From the pore fraction analysis and the observation of the binarized pore maps it can be concluded that the increase in the sintering temperature and longer duration has reduced the fraction and sizes of the pores.

The Gadolinea content was also substantially reduced and the cluster sizes are also much smaller in sample II.

For the analysis of the homogeneity of the remaining matrix, histograms of thermal diffusivity distribution were preformed for each image. In Fig. 2 the two histograms are presented in a scale preserving the area, obtaining-a mean value of $\langle D \rangle= 0.0185 cm^2/s$ for type I and $\langle D \rangle= 0.0194 cm^2/s$ for type II with a standard deviation of $\Delta D= 0.0129 cm^2/s$ and $\Delta D= 0.0046 cm^2/s$ respectively. The mean values are similar, indicating a similar composition but the larger standard deviation clearly indicates a lesser homogeneity for type I.

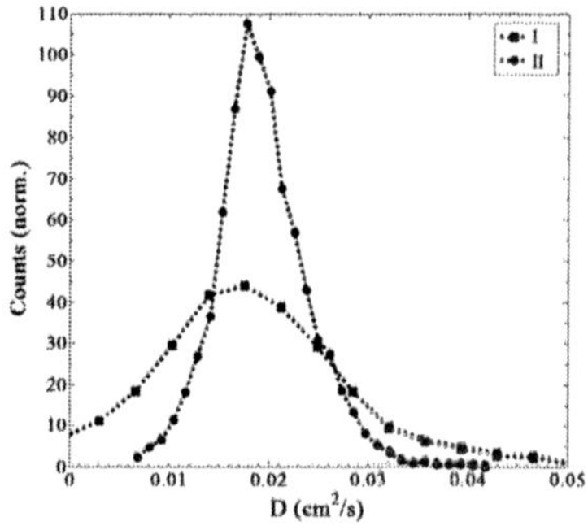

Figure 2. Histograms of the distribution of thermal diffusivity for both samples as obtained from the processed images of figure 1c.

CONCLUSIONS

As the thermal diffusivity of the interdiffused (U, Gd)O_2 phases is much lower than any of the pure oxides, the technique can distinguish inter-diffused material from conglomerates of pure gadolinea and urania nanoparticles, even in the case where the particles are much smaller than the optical resolution of the microscope. In fact such mixtures would have thermal diffusivities larger than that of urania. Such information at microscopic level could only be acquired for similar samples with a delicate and time consuming electron back scattering diffraction experiment.

The technique not only determined the thermal diffusivity but also in the same scan was able to localize, quantify and measure the microscopic and mesoscopic pores providing the relative area they occupy. Similar measurements of gadolinea clusters were performed using the distinct electronic bandgap of gadolinea as compared to urania. Using green pump laser light the urania phases were strongly absorbent, providing a large photothermal signal while gadolinea did not absorb and hence gave a null signal that was used to identify its presence.

Due to the lower thermal diffusivity of interdiffused uranium and gadolinium oxides as compared to pure gadolinea and urania, the technique can establish if the sintering process yielded a homogeneous interdiffused sample or else if there is a mixture of grains. Clusters of urania grains with low or null gadolinium content were also identified in this manner.

The comparison of samples manufactured with different procedures allowed a clear identification of the improvements obtained after changing the fabrication protocol.

Finally it can be stressed that as the technique provides quantitative measures of the homogeneity of the samples at microscopic levels, it provides a protocol that can be used as a reliable specification and quality control method for nuclear fuels.

REFERENCES

1. M. Durazzo , F.B.V. Oliveira , E.F. Urano de Carvalho , H.G. Riella, Phase studies in the UO_2–Gd_2O_3 system, Journal of Nuclear Materials 400 (2010) 183–188
2. TECDOC-844, International Atomic Energy Agency, Characteristics and Use of Urania-Gadolinia Fuels, ISSN 1011-4289, Viena, Austria, 1995.
3. L. Hälldahl, S. Eriksson, Characterization of homogeneity in (U, Gd)O_2-pellets, Journal of Nuclear Materials 153 (1988) 66-70.
4. O.E. Martínez, F. Balzarotti, N. Mingolo, Thermoreflectance and photodeflection combined for microscopic characterization of metallic surfaces, Applied Physics B 90 (2008) 69-77.
5. U. Crossa Archiopoli, N. Mingolo, O. E. Martínez, Two-dimensional mapping of micro-hardness increase on surface treated steel determined by photothermal deflection microscopy, Surface and Coatings Technology, 205 (2011) 3087-3092.
6. U. Crossa Archiopoli, N. Mingolo, O. E. Martínez, Two-dimensional imaging of thermal diffusivity in metals by scanning photodeflection detection, Journal of Applied Physics 107 (2010) 023520-1 to 6.
7. Li Bincheng, J. P. Roger, L. Pottier, and D. Fournier, Complete thermal characterization of film-on-substrate system by modulated thermoreflectance microscopy and multiparameter fitting, J. Appl. Phys. 86, (1999). 5314-1to3.

8. D. Fournier, Thermal-Wave Probing at Various Spatial Scales, MRS Bulletin 26, (2001) 465 -470
9. D. Rochais, H. Le Houëdec, F. Enguehard, J. Jumel, and F. Lepoutre, Microscale thermal characterization at temperatures up to 1000°C by photoreflectance microscopy. Application to the characterization of carbon fibres, J. Phys. D 38, (2005).1498-1503.
10. A. Rosencwaig, J. Opsal, W. L Smith, and D. L Willenborg, Detection of thermal waves through optical reflectance, Appl. Phys. Lett. 46, (1985). 1013-1015.
11. A. Rosencwaig, J. Opsal, and D. L. Willenborg, Thin-Film thickness measurements with thermal waves, Appl. Phys. Lett. 43, (1983). 166-168.
12. Opsal, A. Rosencwaig, D. L. Willenborg, Thermal-wave detection and thin-film thickness measurements with laser beam deflection, Appl. Optics 22 (1983) 3169-3176.
13. M. P. Singh, C. S. Thakur, K. Shalini, S. Banerjee, N. Bhat, S. A. Shivashankar, . Structural, optical, and electrical characterization of gadolinium oxide films deposited by low-pressure metalorganic chemical vapor deposition, J. Appl. Phys. 96 (2004) 5631-5637
14. J.K. Fink, Thermophysical properties of uranium dioxide, Journal of Nuclear Materials 279 (2000) 1-18.
15. M. Amaya, M. Hirai, H. Sakurai, K. Ito, M. Sasaki, T. Nomata, K. Kamimura, R. Iwasaki, Thermal conductivities of irradiated UO_2 and $(U, Gd)O_2$ pellets, Journal of Nuclear Materials 300 (2002) 57–64.

Mater. Res. Soc. Symp. Proc. Vol. 1492 © 2013 Materials Research Society
DOI: 10.1557/opl.2013.557

4K-GM Cryocooler Performance and Thermal Conductivity of $Ho_xEr_{1-x}N$

Takanori Nakano[1], Yusuke Hirayama[1], Takushi Izawa[1], Takashi Nakagawa[1], Yasushi Fujimoto[2], Shinji Masuyama[3], Toshio Irie[4], Eiji Nakamura[4], Takao A. Yamamoto[1]

[1] Graduate School of Engineering, Osaka University, 2-1 Yamadaoka, Suita, Osaka 565-0871, Japan.

[2] Institute of Laser Engineering, Osaka University, 2-6 Yamadaoka, Suita, Osaka 565-0871, Japan

[3] Oshima National College of Maritime Technology, 1091-1 Komatsu, Suo-oshima-cho, Oshima-gun, Yamaguchi 742-2193, Japan.

[4] Santoku Corporation, 4-14-34 Fukaekita-cho, Higashinada-ku, Kobe, Hyogo 658-0013, Japan.

ABSTRACT

$Ho_xEr_{1-x}N$ (x=0.25, 0.5, 0.75) samples were synthesized by nitriding of Ho_xEr_{1-x} alloy bars and their thermal conductivity κ were measured. The measured κ values were comparable to those of stainless steel and Er_3Ni. $Ho_{0.5}Er_{0.5}N$ showed the highest κ of the present three samples. The thermal diffusivity calculated from the κ and the specific heat indicates that $Ho_{0.5}Er_{0.5}N$ is a very promising regenerator material for the cryocoolers. The electrical resistivity ρ was also measured as a function of temperature.

INTRODUCTION

Gifford-McMahon (GM) cryocooler is a typical cryogenic refrigerator [1] widely used for helium and hydrogen liquefaction, and for systems using superconducting magnets such as diagnosis devices based on the magnetic resonance imaging. It generates cryogenic temperatures by the regenerator equipped in the system, and its cooling performance strongly depends on properties of the material packed in the regenerator. Required physical properties are high specific heat, C, and high thermal conductivity, κ, below 15 K, because the cool generated by the expansion of helium gas is desired to be stored and released as much and quickly as possible. In addition, the material should be shaped into spheres with smooth surface, because helium gas needs to flow with the least hydrodynamic resistance through the regenerator column packed with the spheres. Therefore, processing technology for fabricating the sphere is also of great importance.

We have pointed out that rare earth nitrides [2-6] have high specific heats due to their magnetic transitions at cryogenic temperatures and they are regarded as the advanced regenerator material [7,8]. We executed cooling tests with a two-stage GM cryocooler packed with HoN and ErN spheres, and observed that both the materials made temperature below 4.2 K [9,10]. The cooling power, *CP*, obtained by ErN was found to exceed 0.1 W at 4.2 K, by using a commercial GM cryocooler of 0.1-W nominal *CP* [11]. Actually measured *CP* with the ErN sample was slightly smaller than with $HoCu_2$. The Er sample included imperfections such as surface irregularity, pores in the bulk, and a considerable oxygen impurity, nevertheless the ErN sample was in keen competition with $HoCu_2$. Its material processing is, of course, to be improved. We propose another improvement on its physical property by "alloying" of HoN and ErN. The specific heats of HoN and ErN are smaller than that of $HoCu_2$, respectively below 10 K and above 5 K. To obtain large *CP*, the peak of the specific heat should be broad so as to cover the 5-10 K region, as the curve of $HoCu_2$ does by the twin peaks. We have synthesized binary nitrides $Ho_xEr_{1-x}N$ (x=0.25, 0.5, 0.75), and found that the specific heat of $Ho_{0.5}Er_{0.5}N$ is higher than $HoCu_2$ in this region [8]. In this work, we report on the thermal conductivities κ of the $Ho_xEr_{1-x}N$ samples (x=0, 0.5, 0.75, 1). The value of κ influences on the dynamic heat transport in the regenerator, because it is described by the differential equation, $\partial T/\partial t = -D\nabla^2 T$, where $D=\kappa/C$ is thermal diffusivity. It is straightforward that high *D* is preferable for rapid heat transport but a large *C* decreases *D*.

EXPERIMENT

The nitride samples for measurements of thermal conductivity were synthesized by nitriding of alloy bars of Ho_xEr_{1-x} (x =0, 0.5, 0.75, 1) of 2×2×15 mm and 99.9 % purity, which were purchased from Nippon Yttrium Co., Ltd, with a hot isostatic pressing device (O_2-Dr.HIP; Kobelco Co. Ltd.). Nitriding temperature and pressure of nitrogen gas (99.9999% purity) were controlled up to 1823 K and 200 MPa to avoid formation of crack and deformation. Thus obtained sample was crack free and its XRD pattern was measured with a diffractometer (RINT Ultima+; Rigaku Corporation) using Cu-$K\alpha$ radiation. All the samples exhibited the diffraction pattern of the NaCl type crystal structure, indicating formation of single phase material of mononitride. The thermal conductivity was measured in the temperature range of 2 - 300 K using the PPMS (Quantum Design Co., Ltd.) equipped with the Thermal Transport Option (TTO). An amount of heat was allowed to flow into the sample bar from the heater shoe to create a user-specified temperature difference ΔT along the bar with distance Δx. To obtain κ, this system dynamically models the thermal response of the sample to the low-frequency, square-wave heat pulse, then calculates κ from the relation, $\kappa = -q\Delta x / A\Delta T$, where q is heat flux and A is cross-section of the sample bar. The TTO system also provides the four-probe lead configuration which enables to measure the electrical resistivity ρ at the same time.

RESULTS AND DISCUSSION

In figure 1, ln κ measured with $Ho_xEr_{1-x}N$ (x=0, 0.5, 0.75, 1) are plotted against ln T together with those of stainless steel, Er_3Ni and Pb for comparison. It is noticed that all the curves of nitride have their maximum around 100 K and decreases with ln T with an almost common slope, corresponding to $\kappa = AT^{1.6}$ (A is a constant). No significant influence of the magnetic transition is noticed in the curves. These κ values of $Ho_xEr_{1-x}N$ are comparable with those of stainless steel and Er_3Ni [12,13], and that of $Ho_{0.5}Er_{0.5}N$ is the highest of the present nitrides. Figure 2 (a) shows temperature dependence of thermal diffusivity, $D = \kappa / C$. The values of C and κ of nitrides other than those measured in this work were quoted from our previous work [7]. Those of Er_3Ni and $HoCu_2$ were quoted from literature [14]. Since data of $\kappa(HoCu_2)$ was unavairable, the curve of $HoCu_2$ was drawn by substituting κ(stainless steel) instead. In figure 2(b), ln C-ln T curves are shown, which has the transverse axis in common with figure 2(a).

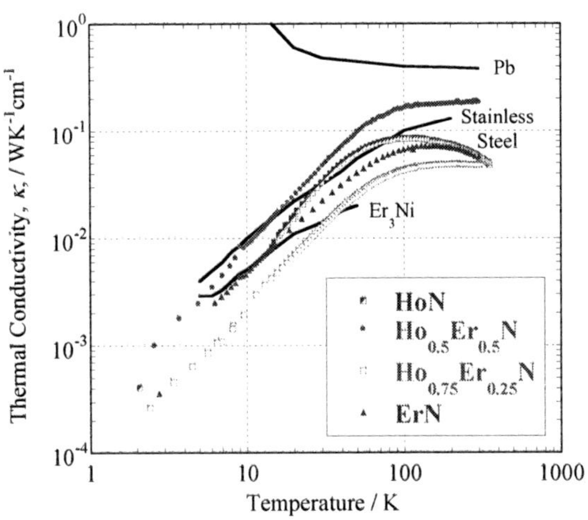

Figure 1. Thermal conductivity of $Ho_xEr_{1-x}N$ (x = 0, 0.5, 0.75, 1)

It is clear that the features of ln C - ln T curves are well reflected on the ln D - ln T curves. The very sharp and high peak on C of HoN brings about a very sharp and deep valley in the D curve. The peaks on C of the other materials are not so sharp enough to make such valley in their D curves, but make steps at their peak temperatures. It is noticed that $Ho_{0.5}Er_{0.5}N$ has the highest D of the binary nitrides and keep high values in the temperature range around and below 10 K which is resirable for the 4 K cryocooler. This D value of $Ho_{0.5}Er_{0.5}N$ is several times higher than that of ErN which has provided the cooling power performance in keen competition with $HoCu_2$. Although the specific heat peak of ErN just around 4 K is significantly high, its low D may have held back the cooling power. Note that D(ErN) is the lowest of the materials displayed there. On the contrary, $Ho_{0.5}Er_{0.5}N$ has both of high D and C in 3-15 K of temperature region. Thus, $Ho_{0.5}Er_{0.5}N$ is the most promising candidate for the advanced regenerator material.

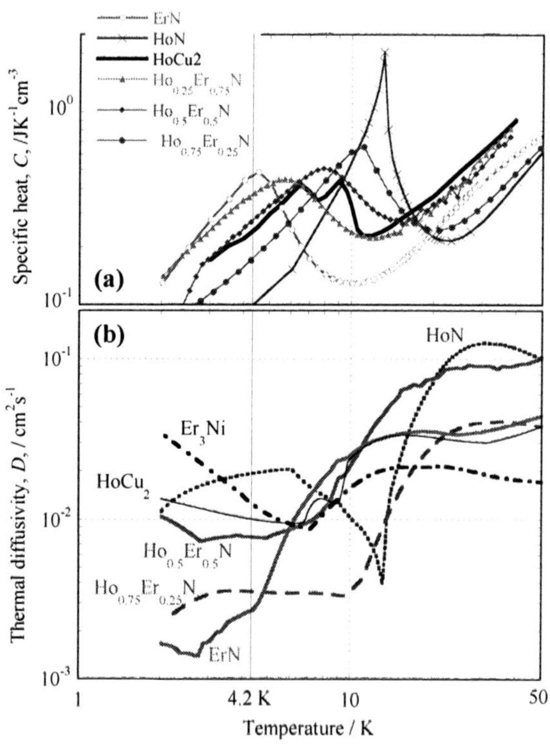

Figure 2. (a) Temperature dependence of volumetric specific heat of the nitride samples and $HoCu_2$. (b) Temperature dependence of their thermal diffusivity D.

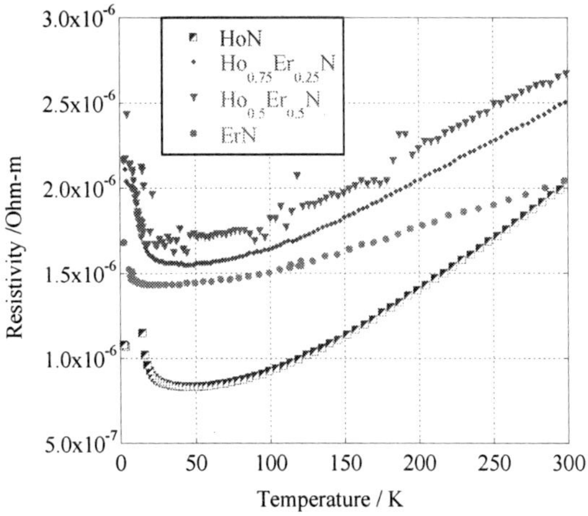

Figure 3. Temperature dependence of the electrical resistivity of $Ho_xEr_{1-x}N$ (x = 0, 0.5, 0.75, 1)

In figure 3 the electrical resistivity ρ of $Ho_xEr_{1-x}N$ (x=0, 0.5, 0.75, 1) are shown. Above 100 K, a general trend was observed that ρ increases linearly with temperature, suggesting these nitrides are metallic. This observation is consistent with the report by Wachter for GdN [15]. The electrical conduction behavior seems sensitive to the stoichiometry of the nitrides, so that further discussion may not meaningful in this stage. In lower temperature range, however, they exhibit increases in ρ with decreasing temperature. The curve of HoN has a clear discontinuous edge at about 15 K, corresponding to its Curie's temperature. Other curves seem to have similar discontinuities, but experimental data is not enough for further discussion.

To summarize, we have synthesized $Ho_xEr_{1-x}N$ samples by the hot isostatic pressing method and measured their thermal conductivities and electrical resistivities. The thermal diffusivities were calculated from these data and our previous data of their specific heat. The results indicated that $Ho_{0.5}Er_{0.5}N$ has better properties than that of ErN from the both viewpoints of specific heat and thermal diffusivity in the temperature range relevant to the regenerator material for the 4K-GM cryocooler. Since we have already demonstrated that ErN gives a cooling power comparable with that of $HoCu_2$, $Ho_{0.5}Er_{0.5}N$ is much more promising candidate material as the advanced regenerator.

REFERENCES

1. A.T.A.M. deWaele, J. Low Temp. Phys. **164**, 179–236 (2011).
2. T. Nakagawa, K. Sako, T. Arakawa, T.A. Yamamoto, J. Alloys and Comp. **364**, 53–58 (2004).
3. T.A. Yamamoto, T. Nakagawa, K. Sako, T. Arakawa, H. Nitani, J. Alloys and Comp. **376**, 17–22 (2004).
4. T. Nakagawa, K. Sako, T. Arakawa, N. Tomioka, T.A. Yamamoto, K. Kamiya, T. Numazawa, J. Alloys and Comp. **408–412**, 187–190 (2006).
5. T. Nakagawa, T. Arakawa, K. Sako, N. Tomioka, T.A. Yamamoto, T. Kusunose, K. Niihara, K. Kamiya, T. Numazawa, J. Alloys and Comp. **408–412**, 191–195 (2006).
6. Y. Hirayama, T. Nakagawa, T. Kusunose, and T.A. Yamamoto, IEEE Trans. Magn. **44** (11), 2997-3000 (2008).
7. S. Nishio, T. Nakagawa, T. Arakawa, N. Tomioka, T.A. Yamamoto, T. Kusunose, K. Niihara, T. Numazawa, K. Kamiya, J. Appl. Phys. **99**, 08K901 (2006).
8. Y. Hirayama, N. Tomioka, S. Nishio, N. Kusunose, T. Nakagawa, K. Kamiya, T. Numazawa, T.A. Yamamoto, J. Alloys and Comp. **462**, L12–L15 (2008).
9. T. Nakano, Y. Hirayama, T. Izawa, T. Nakagawa, T.A. Yamamoto, Y. Fujimoto, S. Masuyama, T. Irie, E. Nakamura, Proc. 24th Int. Cryog. Eng. Conf., May14-18, 2012, Fukuoka, Japan. (in press)
10. T. Nakano, S. Masuyama, Y. Hirayama, T. Izawa, T. Nakagawa, Y. Fujimoto, T. Irie, E. Nakamura, T.A. Yamamoto, "ErN and HoN spherical regenerator materials for 4K-GM Cryocooler," J. Appl. Phys.. (to be accepted)
11. J. Bischof, M. Diviš, P. Svoboda, Z. Smetana, Phys. Stat. Sol. (A) Appl. Res. **114**, K229-K231 (1989).
12. M. Sahashi, Y. Tokai, T. Kuriyama, H. Nakagome, R. Li, M. Ogawa, T. Hashimoto, Adv. Cryog. Eng. **35**, 1175-1182 (1990).
13. T. Kuriyama, R. Hakamada, H. Nakagome , Y. Tokai, M. Sahashi, R. Li, O. Yoshida, K. Matsumoto, T. Hashimoto, Adv. Cryog. Eng. **35**, 1261-1269 (1990).
14. M. Ogawa, R. Li, T. Hashimoto, Cryogenics **31**, 405-410 (1991).
15. P. Wachter, Results in Physics **2**, 90–96 (2012).

Material Resources

Mater. Res. Soc. Symp. Proc. Vol. 1492 © 2013 Materials Research Society
DOI: 10.1557/opl.2013.257

Sustainability in Materials Research in the EU: From FP7 to Horizon 2020

Renzo Tomellini[1], Johan Veiga Benesch[1]
1. DG Research & Innovation, European Commission, 1049 Brussels, Belgium.

ABSTRACT

Environmental issues steadily receive more and more attention at EU policy level. This can for example be seen in the Raw Materials Initiative by DG Enterprise and Resource Efficient Europe by DG Environment which goes back to the theme of a sustainable economy as expressed by the Europe 2020 growth strategy. DG Research and Innovation supports related research activities. The Nanotechnology, Materials & Production (NMP) Theme in the FP7 Cooperation scheme has taken stock of this, by for example including aspects such as substitution, life cycle assessment, improved resource efficiency and better performance materials in the NMP calls for proposals. This is done with the aim to achieve a more green economy and fostering more sustainable consumption and production patterns.

Research on better performing and sustainable materials will more than ever be a pre-condition to meeting such challenges. Progress will come through the development of intelligent materials that embed and transfer knowledge into products and processes or perform certain tasks, when in use or during manufacturing. Already, some 70 per cent of all technical innovations hinge directly or indirectly on the properties of the materials employed. We have passed from the perception "materials are in the drawer" to the perception "materials are the bottleneck". The next step can be "materials are the solution".

At least 60 % of the total proposed Horizon 2020 budget is related to sustainable development, the vast majority of this expenditure contributing to mutually reinforcing climate and environmental objectives. In a resource-scarce Europe, new products must have low material / energy resource needs and high knowledge content. As stated in the Europe 2020 strategy, endorsed by EU leaders: "Europe must promote technologies and production methods that reduce natural resource use, and increase investment in the EU's existing natural assets".

Materials can have a large environmental impact in many of its stages, from sourcing, extraction, processing, auxiliary materials and processes, use and end of life fate. The choice or design of material solutions can thus have a great impact on the technologies in which they are used. Implying that a material could be an integral part of the solution to a problem created by the use of a specific technology. Such solutions could require entirely new materials either to replace a material or be part of a new technology based on better performing materials and eco-designed products.

INTRODUCTION

For many years, industry progressed with the idea that suitable materials were always available, but now we see that this is not the case for a number of challenges we face: sustainable energy production and storage, future ICT, green transport and health applications. We have passed from the perception "materials are in the drawer" to the perception "materials are the bottleneck". The next step can be "materials are the solution".

Research on better performing and sustainable materials will more than ever be a pre-condition to meet such challenges. Progress will come through the development of intelligent materials that embed and transfer knowledge into products and processes or perform certain tasks, when in use or during manufacturing. Already, some 70 per cent of all technical innovations hinge directly or indirectly on the properties of the materials employed [1].

Environmental issues require more and more attention at EU policy level. This can for example be seen in the Raw Materials Initiative by DG Enterprise [2] and Resource Efficient Europe by DG Environment [3] which goes back to the theme of a sustainable economy as expressed by the Europe 2020 growth strategy [4]. DG Research and Innovation supports related research activities, and the Nanosciences, Nanotechnologies, Materials and new Production Technologies (NMP) Theme in the Seventh Framework Programme (FP7) Cooperation scheme has taken stock of this. This would include aspects such as substitution, life cycle assessment, improved resource efficiency and better performance materials in the NMP calls for proposals.

Materials and their use can have a large environmental impact in many aspects, from sourcing, extraction, processing, auxiliary materials and processes, use up to end of life. The choice or design of material solutions can thus have a great impact on the technologies in which they are used. This implies that a material could be an integral part of the solution to a problem created by the use of a specific technology. Such solutions could require entirely new materials either to replace a material or be part of a new technology based on materials that perform better.

In materials science and engineering, research is a long-term investment, but innovation must be as rapid as possible. The success that European industry is experiencing today is largely due to research efforts over the past twenty to thirty years. Future success will depend upon decisions taken today and convinced engagement.

MATERIALS AS ENABLERS

The European Commission has recognised the relevance of advanced materials and singled them out for support in its Key Enabling Technologies (KETs) initiative, launched in 2009 [5]. Based on current global research and market trends the following could be regarded as the most strategically relevant KETs, given their economic potential, contribution to solving societal challenges and knowledge intensity: nanotechnology, micro-nano-electronics, advanced materials, photonics, industrial biotechnology and advanced manufacturing systems.

Mastering and deploying KETs in the European Union is central to strengthening industrial innovation and the development of new products and services needed to deliver smart, sustainable and inclusive European growth. Significant economic benefits can be expected, offering a widening variety of uses in an increasing number of application areas and industries, as well as contributing to energy and resource efficiency, through innovative materials, processes, technologies and applications. In the FP7 theme NMP materials and nanomaterials are thus seen as important enablers for all industrial applications and technologies. These 3 KETs are proposed to receive continued support under the Leadership in Enabling and Industrial Technologies within the Horizon 2020. The potential of these key enabling technologies is largely untapped. Increasingly systemic solutions will need to evolve in order to address major societal challenges, such as ensuring high-speed communication, the environment, finding appropriate transport solutions, and addressing the energy question. Low carbon technologies and applications will play a vital role in reaching European energy and climate change targets. Improvements in the KETs could lead to better resource and energy efficiency. Also, their environmental impact needs to be assessed in a life-cycle perspective, taking advantage of the related initiatives promoted at EU level in this context. For an integrated policy approach to KETs, legitimate health and environmental consequences also need to be actively addressed.

MATERIALS ACROSS BOUNDARIES

Advances in science are also blurring boundaries between disciplines: physics, chemistry, biology. We are aware that, nowadays, substantial innovation may arise from a convergence of disciplines, such as in nanobio (or bionano) technology. We are witnessing a step change in complexity: from mono-disciplinarity to multi-, inter- and supra-disciplinarity. With an increasing world population and finite resources, industry and economies are moving from being resource-based to knowledge-based. To keep improving the European way of life and to create more and better jobs, European companies should receive support to be innovative and stay competitive. Science and technological progress are crucial. But there will also be an increasing need to derive economic benefits from research and innovation by bringing new applications into the marketplace.

Multi-scale materials modelling have tremendously progressed and Europe is leading in this. The capacity to conceive and produce materials by design from an atomic or molecular level is making it possible to more accurately master industrial production. With recent advances in parallel computational power and availability, predictive multi-scale material modelling has the potential to grow stronger with the result of enabling more systematic and faster development of efficient, economically-viable and environmentally-friendly materials, processes and products.

Materials research has an essential role to play in meeting the increasingly complex needs of industry and society, for example greener, smaller, lighter and cheaper products. But while it is essential to examine all aspects of a product's life cycle to make substantial reductions in the use of resources, we also need to address any environmental and health concerns. Failure to do so may slow the take up of innovations.

SUSTAINABLE AND USEFUL MATERIALS

As stated in the Europe 2020 strategy endorsed by EU leaders: Europe will work to "promote technologies and production methods that reduce natural resource use, and increase investment in the EU's existing natural assets" [4]. Many topics under the FP7 theme NMP aim at just that. (See also the following sections on the Raw Materials Initiative, the Materials Roadmap and Private Public Partnerships). Here follow some examples to highlight the attention environmental issues and topics have received in the FP7 NMP theme.

Packaging materials have been focussed on with calls aimed to developed renewably sourced raw materials as alternatives to fossil base, e.g. forestry by-products, recycling and sorting of waste, reduction in the amount of used material, and/or speedier biodegradation.

Sustainable chemistry is a very wide issue that can be seen in a large variety of topics and industrial applications, such as CO2 as feedstock material for the production of fine chemicals, materials for membranes (water purification), materials for photo-catalytic processes for water purification, flexible efficient processing for polymers, nanostructured catalysts with tailor-made functional surfaces and materials for protection, conservation and restoration of cultural heritage assets.

Any "dark side" or drawbacks to new materials must be duly ascertained. Or as the Greek philosopher Epictetus wrote: "materials, of themselves, affect us little; it is the way we use them which influences our lives". For example, the possible toxicological or eco-toxicological effects must be studied according to a precautionary approach. Since the first emergence of nanotechnology in materials engineering, the European Commission has elaborated and promoted a model of integrated, safe and responsible development. There has been a double ethical concern: to prevent risks that may derive from a too-rapid deployment of novel technologies that are not yet known well enough and, at the same time, a desire to allow citizens to benefit from technological advances, as in the case of new treatment of diseases via targeted delivery of nanoparticles.

Depending on the scope of topic in FP7 NMP call for proposal it may be deemed important to include handling of environmental concerns in the proposals. Often the inclusion of such issues is encouraged. In the cases where higher levels of technology readiness is expected these issues are included as requirements to be assessed by the external experts during the evaluations of proposals.. These requirements can be stated in various fashions ranging for statements that environmental concerns such as recycling and end-of-life could be taken into account, activities to handle REACH compliance may be included, or Life Cycle Assessment (LCA) is needed for the process of choosing the optimal material in the selected applications. Admittedly it is complicated and complex to make a thorough Life Cycle Assessment at RTD stages considering that one may encounter; new sources of materials, new material solutions, new production process and new end-of-life options. All of which together may lead various levels on uncertainty in the input data for the LCA.

ENHANCING SUSTAINABILITY IN RESEARCH AND INNOVATION

In our daily lives we rely on materials to perform their functions in the electronics in the mobile phones to wake us up in the morning, in the insulations in the fridge to keep the yoghurt cool, in magnets in the electrical engines in trams to take us to work, in aluminium sheets and high performance ceramics in jet engines in passenger planes that takes us to faraway places, in photovoltaics in satellites to make sure international videoconferences run smoothly. At the same time the number of elements in circuit boards have increased drastically, from about 11 in the 80´s to currently over 45 [6]. So many elements are now constantly put in the techno-sphere, whilst the end of life handling in many cases is limited to dumping or scrapping. Just a few transition elements (e.g. Fe, Al, Cr, Zn etc) have end of life recycling rates over 50%. Meanwhile for many elements (Rare earths, Li, Ga, In etc), used in high tech applications, recycling rates does not even reach 1% [7]. Currently Europe's demand for resources reaches far beyond its borders. Demand for materials is so intense that between 20 and 30 % of the resources we use in Europe are now imported. This overall continuous increase in the use of resources has meant that the ecological footprint of humanity reached one planet in the mid-1980´s [8].

It is thus easy to realise that it is imperative that Europe makes more efficient use of both virgin materials and waste. Does this imply the need for a transition towards a new "socio-ecological" production and consumption model? Tensions between rapidly growing demand and restricted supplies due to the resources available or their polluting nature should cause a constant rise in prices that could be contained by an increased use of renewable resources as well as progress in the reduction of the consumption of the resource. One can call it material productivity, less input materials and less use of materials in a product. That in turn could make a product overall more commercially interesting, hence with increased popularity the total use of a resource might actually increase, and thus one reaches the rebound effect that would seem to cancel out the effect of the efforts to increase the resource efficiency.

In the midst of this, advanced materials that are designed to be multifunctional often lead to complex products that may be complicated to reuse or recycle. At least structurally complex materials have a large risk of being down-cycled if only the materials and not the structure are maintained during recycling. This in turn, should encourage us to take an integrated stand so as to design materials and products for end-of-life. Hence we arrive at the notion of aiming for a circular economy in which waste is a resource [9]. In our current approaches to the flows of materials, products and production there are indeed varied potential for circularity. More importantly, in many case high potentials are far from being captured to their fullest possible extent. There are some glimmers of light; to some extent that these potentials already very well captured in some islands in the techno-sphere were the industrial and community stakeholders go together to form an Industrial Symbiosis. Thus we suggest the final target for research and innovation on materials and materials solutions to be a competitive, green and safe industry.

THE RAW MATERIALS INITIATIVE (RMI)

Many technologies that support sustainable development rely on critical materials. The Raw Materials Initiative started as sounding board based on three pillars; international trade, improved supply of critical resources and boosting resource efficiency. The first report from the RMI defined 14 groups of materials as critical for the European industry and society [10] based on their economic importance and risks in the supply chain. The following discussion has noted that environmental impact might be taken into account in future assessments of criticality. The main driving emerging technologies for the critical raw materials are the following:

Raw material	Emerging technologies (examples from the ad-hoc group)
Antimony	ATO, micro capacitors
Cobalt	Lithium-ion batteries, synthetic fuels
Gallium	Thin layer photovoltaics, IC, WLED
Germanium	Fibre optic cable, IR optical technologies
Indium	Displays, thin layer photovoltaics
Platinum (PGM)	Fuel cells, catalysts
Palladium (PGM)	Catalysts, seawater desalination
Niobium	Micro capacitors, ferroalloys
Neodymium (rare earth)	Permanent magnets, laser technology
Tantalum	Micro capacitors, medical technology

These concepts are included in the evolution of the European Innovation Partnership (EIP) on raw materials, since the supply of raw materials has been identified as key challenge to our society. The proposed Partnership aims to promote innovation along the entire value chain of raw materials. It is thus expected to be a major step for the EU in regaining a major role in securing a sustainable supply of raw materials. This Partnership is designed to bring together Member States and other stakeholders (companies, NGOs, researchers etc.) to develop joint strategies, pull together capital and human resources and ensure the implementation and dissemination of innovative solutions within Europe.

The EIP is a new concept that was introduced in the Europe 2020 flagship Innovation Union. The aim is to address weaknesses, bottlenecks and obstacles in the European research and innovation system that prevent or slow down good ideas being developed and brought to market. EIPs could be used to identify what needs to be done to overcome bottlenecks and galvanise action across public and private sectors. They do not replace funding programmes or regulatory processes, but provide a shared platform for cooperation. With a view to increasing Europe's own production, under the proposal to set up a European Innovation Partnership on raw materials, joint innovation efforts will support exploration, extraction and processing of raw materials. New technologies will help to extract deeper, in more remote areas and under harsh conditions. Action is also needed to develop substitutes for critical raw materials and to improve recycling of electric, electronic equipment and other waste. Improved access to materials will enhance the development of innovative products such as thin photovoltaic layers, energy-efficient lighting, electric cars, advanced passenger jets, infra-red optics and fibre glass.

Some FP7 NMP call topics have been defined in alignment with these concepts, with focus on the two last pillars. Research priorities include mapping, mining (such as unconventional mining and robotics), recycling (including ecodesign), substitution and optimisation in manufacturing. Especially with regards to mining in extreme environments and finding substitutions to critical materials, optimisation of use of critical materials, and also networking activities that could support the EIP on raw materials. In 2011 the call NMP.2011.2.2-4 Novel materials for replacement of critical materials (platinum group metals and rare earths) resulted in three projects currently underway with a joint EU contribution of close to 12M€. The same year the topic NMP.2012.4.1-4 "Substitution of critical raw materials: networking, specifying R&D needs and priorities" aimed for proposals to identify and propose initiatives to help to address relevant EU policies effectively, and also address the specificities of this area such as its industrial, environmental, economic and geopolitical aspects. During its life, the action should collect and elaborate data and make available to the EU Institutions and Member States accurate pictures, analysis of needs, threats and opportunities and proposals for further action. A roadmap of actions, actors and timing, including mapping and networking of existing national activities and/or centres should be developed. The FP7 NMP 2013 call for proposals contains several topics related to mining or substitution and/or optimisation of the use of critical materials. Based on the importance of alignment with EU policies on raw materials and resource efficiency similar and complementary topics can be expected to be published in future calls for proposals.

MATERIALS ROADMAP FOR STRATEGIC ENERGY TECHNOLOGIES

The Strategic Energy Technology-Plan (SET-Plan) was created to establish a low carbon energy technology policy for Europe. Its aim is to accelerate the development and deployment of cost-effective low carbon technologies. The plan comprises measures relating to planning, implementation, resources and international cooperation in the field of energy technology. What this means in practice for EU materials R&D&I is set out in the Commission's recently published "Materials Roadmap Enabling Low Carbon Energy Technologies" [11]. The document, based on contributions from nearly one hundred leading scientists and industrial researchers, show that materials are at the core of developments in the energy technologies covered in the Materials Roadmap: wind energy, photovoltaic energy, concentrating solar power, geothermal energy, electricity grids, electricity storage, bioenergy, carbon capture and storage, nuclear fission, hydrogen and fuel cells, and energy efficient materials for buildings. The Roadmap focuses on material research and development and innovation activities for these low-carbon energy technologies for the next 10 years with market implementation horizons for both 2020/2030 and 2050. It is designed to support coherent and strategic decisions on materials research in industry and academia, both at European level and in EU member states, based on the best available information today.The work might be expanded to cover other technologies as well.

Thus the roadmaps include a large number of material classes and processing activities. These may be designed to address the specific needs of one technology. However, several materials are common to more than one technology. In addition, a broad range of proposed activities have many similarities, such as research and industrial capacities being developed to understand the fundamental nature of materials or processing technologies. Identifying synergies between these activities is of critical importance for the implementation of the roadmap. This could aid in realising economies of scale and scope. It is also expected that cross-technology practises can be pooled together at European level to accelerate the development and integration of innovative materials into low carbon energy technologies.

Many cross-cutting activities have been identified. Standardisation of and common regulations for material specifications, products and testing procedures need to be further promoted to ensure the access of European products to the European and world markets. The importance of information and communication to end-users was highlighted, e.g. via labelling schemes to highlight the benefits of the new low carbon technologies. Supply of critical raw materials since many low carbon energy technologies rely on critical raw materials, such as rare-earth metals for which Europe by and large is dependent on imports. Also, activities related to assessing the sustainability of materials, including the impact of their production, use, recycling, disposal or substitution on a broad range of resources in line with the approach of the Roadmap to a resource efficient Europe [12], are essential for the further development of low carbon energy technologies. To stimulate the market up-take of sustainable and resource efficient materials and technologies requires a conducive environment.

PUBLIC PRIVATE PARTNERSHIPS

Three Public-Private Partnerships (PPPs) were launched as funding tools within the Theme NMP in 2009 as part of the European Economic Recovery Plan [13] to tackle the consequences of the global economic downturn. They have two aims: 1) to boost competitiveness and support employment in three key industrial sectors respectively; manufacturing, construction and automotive, 2) significantly contribute towards a more green and sustainable economy. As consequence of the latter, these topics all have many elements of sustainability very much integrated, both in the call topics and the submitted proposals. In the call for proposals published in 2012 the budget for these covered almost half of the NMP theme.

Factories of the Future have the objective to develop enabling technologies to support EU manufacturing across many sectors. Many calls include aspects such as optimisation of processes to improve resources efficiency both for the product and all its production steps. This includes in-line monitoring of process parameters.

Energy Efficient Buildings with the objective to reduce energy use and CO2 emissions by developing novel building and district concepts, and to promote green technologies and the development of energy-efficient systems and materials in new and renovated buildings. The calls topics have aimed to reduce the overall use of materials, especially fossil based and find renewable alternatives, and also to develop and optimise new processes to build and design building components, whilst also considering endurance and end-of-life aspects.

Green Cars with the objective to support the emergence of more efficient and sustainable hybrid and electric cars, and also to achieve breakthroughs in the use of renewable and non-polluting energy sources, safety and traffic fluidity. Most of the calls for proposal have a focus on battery materials and components, but also light weight structural components.

ENVIRONMENTAL ISSUES IN RUNNING EU FUNDED MATERIALS PROJECTS

The Materials unit of the Directorate for Industrial Technologies of the DG of Research and Innovation recently made an attempt to understand better how project consortia view environmental issues. This would refer to how they could be included and handled in FP7 NMP projects on Materials. A brief questionnaire was devised, comprising 6 questions covering the level of importance of environmental issues with focus on RTD projects, how they are incorporated and affect the outcome of the project as well as including recommendations on how to best handle environmental issues in RTD projects. Environmental issues for the purpose of this questionnaire include, but are not restricted to, resource efficiency, recycling, precautionary and substitution principles, eco-toxicity, bioaccumulation, life cycle assessment, end of life use, sustainable use, green production, and European environmental policies [14].

61 out of the contacted 103 FP7 NMP projects on materials responded of which 74% claimed to include environmental issues. Of those that stated that they do not handle environmental issues most claimed they were of too fundamental a character to include any such issues. Some of those also stated that their outputs have the potential to lead to improved environmental impact by development of better performing material's solutions.

Two main advantages to include environmental issues were identified by the respondents, namely; to develop less harmful materials and/or processes (by substitution or optimisation), and to make the output from projects more attractive to industry. An improved environmental impact could then be demonstrated, such as environmental protection for a secure and healthy life by avoiding harm to people and environment. About half of the respondents use recognised environmental protocols, such as Life Cycle Assessment according to ISO 14040 and/or ILCD handbook, environmental management systems or activities to comply with REACH. It is generally expected that newly developed materials would be correlated to environmental friendliness, which can be increased by using the feedback from life cycle assessment support during its development. Some noted the importance of handling environmental, health and safety (EHS) in conjunction.

Almost two thirds say that inclusion of environmental concerns help to steer the output from the projects, such as helping in selecting optimal materials or processes. Approximately one third said environmental issues are the basis for the projects. Overall, consideration of environmental issues helps to create a technological culture that avoids future human and environmental damage, which is costly in human and economic terms. It was also perceived to increase the need to develop new (environmentally friendly) materials and substitute materials. This would encourage industry towards innovation and exploitation resulting in technological leadership of European industries. For many of the respondents a big justification for the project was to develop new environmentally friendly materials solutions, e.g. sustainable sources materials for products based on eco-design and/or with entire life cycle in mind. Better environmental behaviour is expected to improve the competitiveness of new materials and their market value.

Tackling environmental and eco-toxicity issues upstream in the innovation value chain is by the respondents overall seen as advantageous, as it allows avoid investing in research in technologies which finally would not go into industrial production, and to pave the way to accelerate industrial take-up of promising material solutions.

The respondents' highlighted the value also of on non-monetary support from the European Commission such as networking between projects and between projects and other stakeholders. It was suggested to explore the possibilities for cross-project activities including data and tools sharing and horizon mapping. The setting up of a tight network among projects related to the various environmental issues was seen as desirable. Further suggestions entailed training (e.g. LCA, REACH, EU regulations, work place safety etc.), environmental guidelines (for proposers and running projects) and EU database (single entry point) on environmental issues (e.g. guidelines and regulations). "

The most common recommendation from respondents for projects on materials was to include environmental issues as soon as possible to get an early view of the required research efforts and regulatory requirements so as to be able to adapt the projects accordingly, while at the same time, not losing valuable time on less sound material options. Among the suggestions was noted the need to define environmental requirements, risks and specifications at the beginning of the project, and to analyse the most recent status of environmental regulation. The respondents also recommended to include pertinent expertise, use recognised protocols and to look at the entire life cycle of the studied materials.

Identification of the required specialised fields of expertise needed is essential and, as a consequence, the inclusion of personnel already qualified or ready to undergo specific trainings on handling the relevant environmental issues. Assessment of environmental issues in a life cycle perspective, in combination with comprehensive safety considerations and risk assessment of processes, is considered as crucial in order to optimise the environmental performance and commercialisation of the material technologies under development. Some suggested reinforcing the importance of environmental issues in EU calls for proposals and subsequent evaluations.

HORIZON 2020

Horizon 2020 is the proposed financial instrument to implement the Innovation Union, a Europe 2020 flagship initiative aimed at securing Europe's global competitiveness. Running from 2014 to 2020 with an €80 billion budget, the EU's new programme for research and innovation is part of the drive to create new growth and jobs in Europe. Horizon 2020 should provide funding to help address major concerns shared by all Europeans such as climate change, developing sustainable transport and mobility, making renewable energy more affordable, ensuring food safety and security, or coping with the challenge of an ageing population.

At least 60 % of the total proposed Horizon 2020 budget is related to sustainable development, the vast majority of this expenditure contributing to mutually reinforcing climate and environmental objectives [15]. In a resource-scarce Europe, new products must have low material / energy resource needs and high knowledge content. In the proposal for the European Union's next multi-annual research programme, Horizon 2020, some €13.8 billion are proposed to be dedicated to research and innovation on leadership in enabling and industrial technologies (including KETs). This is expected to lay basis for competitive advantages across a range of sectors and is important for exploitation of synergies and complementarities.

Advanced materials are identified as one of the Key Enabling Technologies in the proposal for Horizon 2020 [16] and several of their suggested issues to be taken into account relate to sustainability. Such as, materials for a sustainable industry with the aim to develop new products and applications and consumer behaviour that reduce energy demand and facilitate low-carbon production, as well as process intensification, recycling, depollution and high added-value materials from waste and remanufacture. Also, optimisation of the use of materials is highlighted with focus on research and development to investigate alternatives to the use of materials and innovative business model approaches. Under Societal Challenges there is also focus on a sustainable supply of raw materials including the need for longer use, high-quality recycling and recovery, and the need to drastically reduce resource wastage. A full life-cycle approach will be taken, from the supply of available raw materials to end of life, with minimum energy and resources requirements. Materials research is also expected to continue to receive funding with the bottom-up schemes, i.e. European Research Council and Marie Skłowdowska-Curie Actions.

For more information on Horizon 2020 see the article "Funding Research and Innovation in the European Union" by the same authors in this issue.

CONCLUSIONS

As stated in the Europe 2020 strategy endorsed by EU leaders: Europe will work to "promote technologies and production methods that reduce natural resource use, and increase investment in the EU's existing natural assets". Sustainability issues are thus gaining in importance for defining research and innovation priorities, from EU policy level down to the operational level of selecting submitted research proposals.

DISCLAIMER

The views expressed in this article are those of the Authors and not necessarily those of the European Commission. This article does not commit the European Commission in any manner. © European Union, 2012. The Materials Research Society has been granted the right to publish it in these proceedings.

ACKNOWLEDGEMENT

Opportunities for European funding of research and innovation can be found on the Participant Portal: http://ec.europa.eu/research/participants/portal/

REFERENCES

1. Die Hightech-Strategie für Deutschland – Erster Fortschrittsbericht, BMBF, 2007
2. COM(2008) 699 final "The raw materials initiative — meeting our critical needs for growth and jobs in Europe"
3. COM(2011) 21 "A resource-efficient Europe – Flagship initiative under the Europe 2020 Strategy"
4. COM(2010) 2020 final "A strategy for smart, sustainable and inclusive growth"
5. COM(2009)512 Final "Preparing for our future: Developing a common strategy for key enabling technologies in the EU"
6. T. McManus, Intel Corp 2006. Cited by Magnus Ericsson in presentation at "Trade and Raw Materials – Looking Ahead", Brussels, 29 Sept 2008 (unpublished)
7. Metals Recycling Rates, UNEP, ISBN 978-92-807-3161-3, 2011
8. Geoffrey de Schutter, WWF. Presented at "Sustainable development: a challenge for European Research" in Brussels, May 2009 (unpublished)
9. Towards Circular Economy: Economic and business rationale for an accelerated transition Ellen Macarthur Foundation, 2012
10. Critical raw materials for the EU, Report of the Ad-hoc Working Group on defining critical raw materials, European Commission, June 2010
11. SEC(2011) 1609 final "Materials Roadmap Enabling Low Carbon Energy Technologies"
12. COM(2011) 571 final Communication from the Commission to the European Parliament, the Council, The European Economic and Social Committee and the Committee of the Regions : Roadmap to a Resource Efficient Europe
13. COM(2008) 800 final "A European Economic Recovery Plan"
14. Johan Veiga Benesch (ed) Report on Responses to Questionnaire on Environmental Issues in FP7 NMP Projects on Materials, DG RTD, EUR 12366 EN, 2012
15. COM(2011) 808 final "Horizon 2020 - The Framework Programme for Research and Innovation"
16. COM(2011) 811 final Proposal for a Council Decision establishing the Specific Programme Implementing Horizon 2020 - The Framework Programme for Research and Innovation (2014-2020)

Mater. Res. Soc. Symp. Proc. Vol. 1492 © 2013 Materials Research Society
DOI: 10.1557/opl.2013.151

Glass: an old material for the future of manufacturing

Susanne Klein[1], Fraser Dickin[1], Guy Adams[1], Steve Simske[2]
[1] HP Labs, Long Down Avenue, Bristol BS34 8QZ, UK
[2]HP Labs, 3404 E Harmony Rd, Ft. Collins, Colorado, 80528-9544, USA

ABSTRACT

Traditional assembly line manufacturing is speculative, costly and environmentally unsustainable. It is speculative because it commits substantial resources—energy, materials, shipping, handling, stocking and displaying—without a guaranteed sale. It is costly because each of these resources—material, process, people and place—involves expense not encountered when a product is manufactured at the time of sale. It is environmentally unsustainable because, no matter how much recycling is done, not using the resources unless actually needed is always a better path. Three-dimensional printing is currently of great commercial interest as it can be employed to manufacture parts on-demand economically and without the significant cost & environmental downsides, i.e. inventory and waste, associated with traditional manufacturing processes. Herein, we describe the formulation of a novel water-based material which can be used in a traditional 3D printer extrusion process to create optically transparent glass-based objects. Such objects have a wide range of applications including, but not wholly limited to: security printing using color & coating effects, protective films and coatings, electronic codes readable by smartphones, tablets or touch screens. Additional all glass objects traditionally manufactured by the so called kiln glass method can be generated by this type of 3D printing making it interesting for the high end market of art objects.

INTRODUCTION

Even after 30 years, the 3D printer market is still a niche market for prototyping and the major obstacle inhibiting its market expansion is the limited number of materials which can be printed and printed together. At present, 3D printing is mostly based on organic materials such as ABS plastics or UV curable monomers. The functionality of the printed objects is constrained by that of the underlying material, e.g. most printed plastic parts are not robust enough to withstand daily usage. Consequently, the main application of 3D printing has been prototyping, and as such it has not yet reached additive manufacturing. To transform 3D printing into a viable method for additive manufacturing, it is necessary to expand the material choice.

The Ragnarok (**R**esearch on **A**dvancing **G**lass & **N**onorganic **A**pplications to **R**ecreate **O**bjects & **K**inetics) project in HP Labs is targeting the more obvious of the replaceable manufacturing processes using 3D printing methods. Casting, molding and sintering are the first three processes receiving our attention.

We identified glass as a promising material for additive manufacturing. Glass is a silica-based material and with 90% of the Earth's crust composed of silicate minerals there is no shortage of silica resources. Glass is easy to recycle and is environmental friendly. With regards to health and safety, inhaling glass dust without the appropriate EHS equipment can lead to silicosis, but the consumer does not usually come into contact with glass dust. Otherwise glass shows no ill health effects. Glass is a cheap material but looks precious, is pleasant to the touch and it's behavioral characteristics are well known such that customers will intuitively know what applications it is most appropriate for.

Glass 3D printing has been attempted in the recent past but the results published [1] lack both the smooth appearance and the optical transparency of traditionally made glass. A photograph showing an example of glass printing by a group in Washington University is shown in Figure 1 below.

Figure 1: 3D printed glass using a powder bed method and sugar as binder. The objects do not look like glass as a result of the powder bed method and the resulting matt surface. There is also discoloration caused by caramelization of the binder material

EXPERIMENTAL DETAILS

To stay as environmentally friendly as possible and to avoid health and safety issues during production, we concentrated on water as the solvent and bio-polymers as binders for our glass pastes. We used binders from the following groups: saccharides, clays, collagens and cellulose. Glass powder, binder and water were worked into a paste which was then extruded through an Auger print head mounted in a RapMan 3D printer, as shown in Figure 2.

Figure 2: Printing of glass paste with a modified RapMan 3D printer

After printing, the objects were thoroughly dried and the so called greenware was then fired in a kiln.

DISCUSSION

Printing by extrusion generates a glossy, that is glassy, surface appearance since the printed object does not come into contact with any powder. Reduction of binder content prevents caramelization. Between 110□C and 180□C all saccharides transform into polymeric caramel. Above 250□C caramel decomposes into carbon mon- and di-oxide, hydrocarbons, alcohols, aldehydes, ketones and several furan derivatives (for more detail see [2]) which are volatile and evaporate from the samples as long as they are not trapped or do not undergo chemical reactions with the glass surface. We have found that reducing the saccharide content of the glass paste to 5 to 10wt% on the weight of glass frit resolves the problem of discoloration. In table 1 we show the minimum binder content for a printable glass paste as a function of the type of saccharide and the resulting discoloration. The required binder percentage for cellulose, collagen and clay binders is on average below 10 wt% and discoloration has not been observed.

Saccharide	Minimum percentage on weight of glass powder	Discoloration
Dextrin	40 wt%	Yes
Dextrin/fructose	25/25 wt%	Yes
Pectin from citrus peel	10 wt%	No
Chitosan	5 wt%	No
Xanthan gum	5wt%	No

Table1: Minimum saccharide content for printable pastes.

In addition, extrusion printing allows the combination of different colors in one printing process, as can be seen in Figure 3.

Figure 3: multiple color glass object.

From a traditional glass production process, kiln glass, it is known that the finer the powder, the finer the detail, but the less transparent the final object. We tested powders with grain sizes in the range from 200 nm up to 200 µm. We have found that glass powders with grain sizes larger than 38 µm samples are translucent. Powders with smaller grain sizes lead to opaque samples (figure 4).

150 µm < diameter < 250 µm translucent	75µm < diameter < 90 µm translucent	38 µm < diameter < 75 µm translucent	Average particle diameter < 38 µm opaque	Average particle diameter: 300nm opaque

Figure 4: Optical appearance as a function of particle size.

The shift in color indicates that the opaque appearance is caused by scattering from glass/air interfaces, but density measurements did not confirm this assumption. The spread of densities within sample groups of the same size is as big as the overall spread over all sizes, as can be seen in figure 5. All samples were processed following the same procedure and firing schedule.

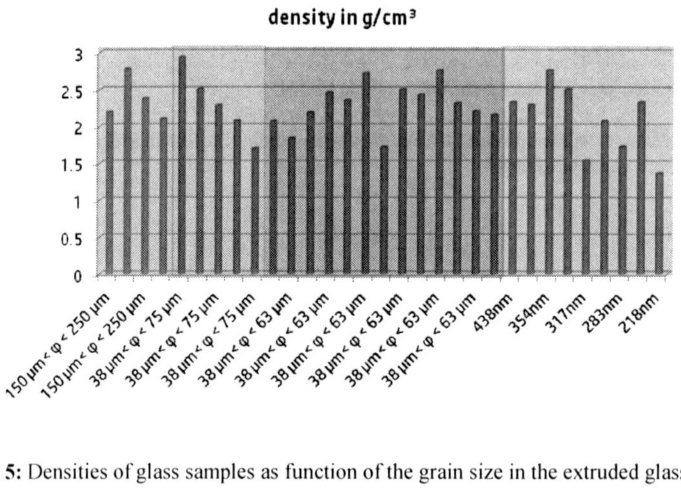

Figure 5: Densities of glass samples as function of the grain size in the extruded glass paste.

CONCLUSIONS

We have developed several formulations for printable glass pastes which were tested with a modified RapMan 3D printer. The best recipes lead to translucent objects which are shape and color true and have the appearance of classic glass. True optical transparency has not yet been reached. Experiments with submicron grain sizes have led to opaque objects with very fine detail, but lacking the required glass-like finish (see also [3]). More research is needed to explain that kind of behavior. To reduce the production time, we have also started exploring paste formulations which allow to the firing step to be skipped without adversely compromising the mechanical integrity of the object. As can be seen in Figure 6, the firing step is the most time-consuming aspect of the process.

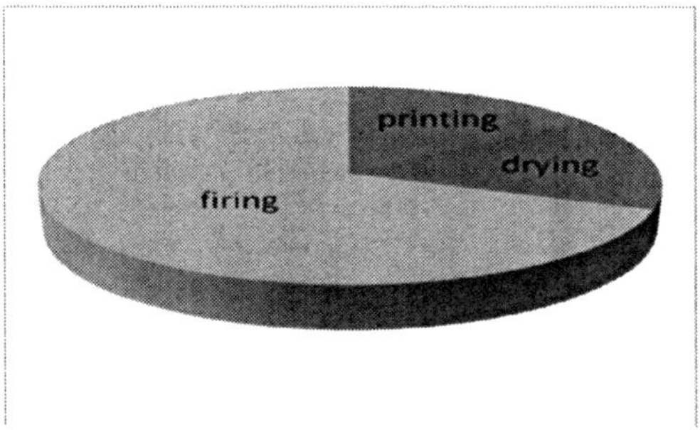

Figure 6: For fired glass, fusing and cooling (summarized as firing shown in green) is the most time consuming part of the production process. Skipping the firing stage would reduce production time radically.

First results show that greenware with cellulose binders show enough structural stability to replace fired glass in certain applications where the optical appearance is not the most important consideration.

REFERENCES

[1] http://open3dp.me.washington.edu/2009/09/3dp-glass-recipe/
[2] J. Claude, J. Ubbink, *Food Chemistry* **96**, 402-410 (2006)
[3] S. Klein, et al., PG 336, *Proceedings of NIP and Digital Fabrication 2012*

Mater. Res. Soc. Symp. Proc. Vol. 1492 © 2013 Materials Research Society
DOI: 10.1557/opl.2013.370

Ecodesign concept case studies: Cu in electric motor and Ni in waste incinerator

Tarja Laitinen[1], Erja Turunen[2], Olli Salmi[2], Kari Tammi[2], Ulla-Maija Mroueh[2], Päivi Kivikytö-Reponen[1]
[1]VTT Technical Research Centre of Finland, PO Box 1300
FI-33101 Tampere, Finland
[2]VTT Technical Research Centre of Finland, PO Box 1000
FI-02044 VTT (Espoo), Finland

ABSTRACT

VTT has implemented the demand of energy and resource efficiency in the framework of Ecodesign concept covering the whole material life cycle from material sources to material design and manufacturing, component life time optimisation and finally recycling concepts. The vision of the virtually supported Ecodesign concept is to create optimized and efficient machine and device components regarding their whole lifecycle by evolving multiscale modelling.

In this presentation we will introduce our development work within our Ecodesign concept by giving two case examples including Cu flow in electrical motor and Ni flow in waste incinerator. In the first case we will discuss raw material scarcity based design criteria, technological challenges and possibilities of Cu substitution and finally energy efficiency in system level. In the latter we will discuss multiscale modelling approach starting from raw materials and new design criteria regarding performance, life time, maintenance strategies and energy efficiency in system level operation.

INTRODUCTION

Energy efficiency is an established, global megatrend throughout the industries in order to secure energy supply, cut costs and mitigate greenhouse gas emissions. Resource efficiency is becoming the demand of industrial activity in sectors like construction, chemicals, automotive, aerospace, machinery or commodities. The alternatives to tackle the raw materials scarcity are either new primary (arctic, seabed) or secondary sources (reuse, recycling) or substitution. Current amount of metals (e.g. Cu, Zn, Ni and Al) that can be directed to recycling from their present use is inadequate compared with the market need. Substitution can be used to develop alternative materials in certain applications or to replace those applications by an equivalent technology that does not rely on the key raw materials.

The typical material life cycle starts from primary production, mining. The ore is converted into metallic form, alloyed with other elements and semi-fabricated in a form of, e.g., powder, sheet or forging for manufacturing the final product. Once the product has served for its purpose, it is either recycled or wasted. The resulting carbon and water footprints are large and solutions are looked for from new technology breakthroughs including new material design concepts, additive manufacturing methods and new recycling and recovery technologies. The future society is based on sustainable development, where the re-use is of central importance. Materials reuse, efficient use of secondary flows as well as more efficient recovery technologies provide the basis for the sustainable exploitation of natural resources.

The past decade has witnessed a sharp increase in the demand and price of metals and other mineral raw materials. This demand increase has its roots in the rapid growth of countries such as China, India and Brazil, where per-capita material inputs into society have experienced up to a seven-fold increase since the early 1970s [1].Urbanization in these countries is considered a key factor in demand increase for bulk metals such as iron, copper and nickel. The case example metals used in this article, Cu and Ni, have experienced a demand growth from 15 to 20 Mt and 1,1 to 1,6 Mt, respectively, during the past decade [2]. This trend is expected to continue. By 2027, global Cu demand may have exceeded 30 Mt [2].

In the wake of the metals demand balance shifting towards Asia, the EU has begun to take its own measures on safeguarding sustainable raw materials availability. The recent report on critical raw materials ranks 41 minerals and metals in order of strategic importance on the basis of supply risk and economic importance [3]. In this categorization, Cu and Ni are considered economically important but not under supply risk. Metals often alloyed with bulk metals, such as rare earths, niobium, and tantalum are, on the contrary, considered critical.

Increasing scarcity of technology metals and high price of bulk metals poses challenges for design and manufacturing. On the positive side, this is not the first time designers are faced with similar challenges. The ecodesign paradigm, dating back to the mid-1990s, emerged as a response to increasing environmental impacts of products and processes, as well as to the difficulties in recycling ever more complex combinations of raw materials in products. "Design for the environment" and "design for disassembly" are examples of concepts belonging to the overall ecodesign paradigm [4]. These concepts become useful as we look at the minerals conundrum of the coming decades. How to design products and processes that are easy to recycle, whose materials do not rank high on supply risk, and whose environmental impacts are smaller than those of the current generation of manufactured products ? How to ensure optimum performance with the variable material challenges ? How to sustain and strengthen local manufacturing also elsewhere than in labour intensive countries ? And not least, how to turn all this into profitable business ?

In this paper, we propose an early attempt to tackle the newly emerged, scarcity-driven, comprehensive design problem. We present an ecodesign framework of virtual, multiscale modelling as a way towards conserving raw materials and energy. The vision is to create optimized and efficient machine components regarding materials, design, manufacturing, performance, service and recycling. We demonstrate our approach through two technology development case studies: one on future electric car engines, and the other on innovative coating solution for power plant boilers.

THEORY/ EXPERIMENTAL

Ni and Cu are amongst the most intensively recycled materials globally. In Europe, 41 % of the demand of Cu and globally about one third of the demand of both Cu and Ni is met by recycling [5],[6]. On the other hand, the United Nations Environment Programme (UNEP) has estimated that on global level the end-of-life recycling rates of base metals, including Cu and Ni are above 50 %. There is still space for improvement of collection and for reduction of losses in the treatment chain. For example, about 65 % of end-of-life Ni is collected for recycling, but only 52 % is recycled into products [5]

The most evident reason for relatively high Cu and Ni recycling rates, especially in developed countries, is the economic profitability. The prices of secondary Ni and Cu are almost as high as the prices of primary metals, and recycling costs are low compared with those of primary production. Also environmental impacts of secondary production are lower. For example, energy savings and greenhouse gas emission reductions are considerable. Because of the long history of Cu and Ni recycling, there is also an existing infrastructure for recycling.

Cu is widely used as pure metal in electrical wires, electronics in general, plumbing and construction materials. Their collection and separation from other materials is relatively easy. When Cu is alloyed with other materials the most profitable option is to recycle it back to alloy production. Cu may also cause some challenges in the production of secondary materials. Separation of the short pieces or fine particles produced from Cu wire in the crushing processes is difficult, and they may deteriorate the quality of the residues and even the product. Ni is mostly used in stainless steel and other metal alloys which can be recycled either as the same alloy, or if mixed scrap is used as a raw material, an alloy of specified composition can be produced. Ni is also used in rechargeable batteries, electroplating and chemicals. The collection and recycling rates of rechargeable batteries are still quite low. Although the recycling rates of Cu and Ni are reasonably high, secondary production cannot cover the growing demand.

Increasing demand for sustainable and efficient machine components regarding materials, design, manufacturing, performance, service and recycling can be answered by ecodesign framework of virtual, multiscale modelling, Figure 1. Multiscale modelling provides a cost-effective and fast tool for iterating and optimising the component material properties or technological performance as a part of a system when compared to prototype building and actual on-site testing. Varying future scenarios can be simulated and verified, a profound understanding of multiscale and multidimensional problems achieved and tools for decision making developed.

Ecodesign concept is a toolbox under development taking into consideration the total product lifetime path. The modelling methods developed must provide an integrated solution starting from material level to component level and finally ending up to system level. In previous research, process simulation models have been applied to green energy supply synthesis under uncertain demand conditions, to tackling waste solvents in chemical industry, as well as to the composition of metal alloys produced from scrap metal [7]. The ecodesign concept is examined through two case examples, namely Cu flow through electric motor and Ni flow through combustion boiler coating.

RESULTS

Case 1 – Cu in electric motors

In 2010, the global amount of passenger cars was more than 1 billion, a number estimated to double by 2020 [8]. Environmental awareness, political decisions to reduce carbon dioxide levels and dependence on fossil fuels, rising energy prices and advances in technology, especially in battery and information technology, are motivating car manufacturers to current large-scale launch of electric vehicles (EV). Technological change puts pressure on scarce materials usage. Primarily, the pressure is on rare earth materials used for instance in permanent magnets and batteries, but a similar pressure is expected to emerge on Cu. An average car with an internal combustion engine contains about 25 kg of Cu of which more than two-thirds is in the car's wiring harnesses and electrical components [9]. A plug-in hybrid EV contains, on average, 74 kg of Cu, with the electric motor contributing to more than 60% of that weight[10].

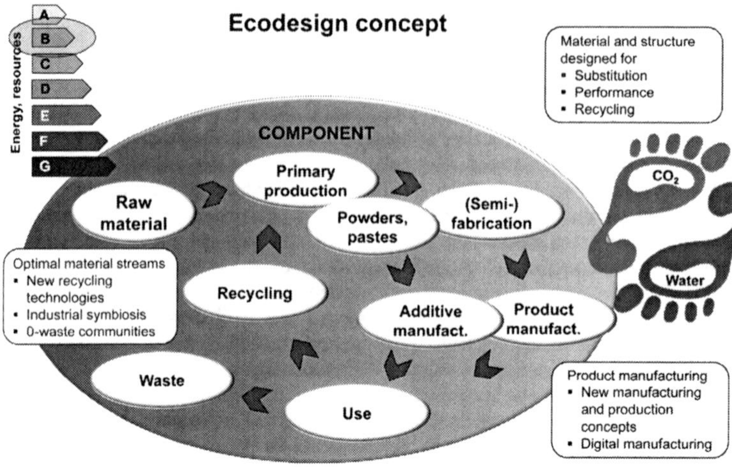

Figure 1. The ecodesign concept.

The market penetration of EVs is expected to intensify between 2020 and 2030. According to a recent estimate, the share of EVs in the EU27 passenger car fleet could increase from a mere 1% in 2020 up to 33% in 2030 [11]. This would imply a significant additional burden on the world Cu resources.

Increasing price of raw materials and the looming raw material scarcity have forced electrical machine researches to find alternative solutions in motor design. Switched reluctance motors have been re-invented during last ten years to reduce the permanent magnet usage [12,[13]. Similarly, alternatives for Cu are being sought, which is more challenging because of the uniquely high conductivity of Cu. The winding resistance is a significant source of losses in electrical machines, especially for low and medium speed machines. Al has replaced Cu in power lines, some power transformers and squirrel cage rotors as techno-economical calculations have shown that the usage of Al is a cost-efficient solution. Aluminum has price and weight benefits, but the conductivity of aluminum is about 60 % of the conductivity of Cu and thus a conductor cross section has to grow 65 % to achieve the same linear conductivity.

For electrical motors, the situation is challenging, because the motor size is usually minimised. Replacing Cu with aluminum has been studied, but the motor sizes tend to grow by 60 %. It has actually been shown that the replacement increases the size of the induction motor and makes the electrical motor with aluminum windings more expensive than trational Cu wound motors [14]. Recently, nano-materials have been investigated for mane applications. Carbon nanotube based cables and wires seem to have multiple good characteristics and means to improve their conductivity have been found, albeit below that of Cu [15]. For the future carbon nanotubes or fibre seems to be ideal solution due to specific weight and reasonable conductivity.

Case 2 – Ni in waste incinerators

The structural and protective materials of the energy production equipment are defining a boundary for the efficiency, competitive position and sustainability for a given process and design. The drive for reduced emissions in thermal power generation will focus the development towards firing and co-firing with biomass and residual waste fuels. The effort will call for improved modelling to predict the properties, integrity and expected life of materials and components for the new and increasingly challenging operational conditions.

Typical waste incinerator environment is extremely corrosive containing app 15000 ppm chlorine. Currently the combustion temperatures cannot be raised over 540 °C because of heavy boiler corrosion. Critical boiler surfaces are weld overlaid with Inconel 625 the lifetime of the overlay being max 12 months in the prevailing conditions. Inconel 625 is Ni based superalloy with alloying elements Cr, Mo, Fe and Nb + Ta. The price of Ni fluctuates affecting heavily on global manufacturing industry, Nb and Ta are both on the list of defined critical raw materials in EU. If superheater temperatures could be raised by 20 °C, the energy efficiency would increase roughly by 1 % and NOX and particulate matter emissions would decrease. The push for higher efficiencies and reduced emissions at competitive cost will call for new solutions but also optimised tuning and application of existing materials.

VTT has developed a new spray coating capturing chlorine and thus preventing corrosion [16]. The laboratory results show better resistance to chlorine than any other conventional method. The coating can markedly increase the service life of boilers and reduce their maintenance costs, enable temperature raise and thus better efficiency and lower emissions, and provide a substitution solution for the current coating with critical raw materials Ni, Nb, Ta.

CONCLUSIONS

The aim of this paper was to present initial steps in tackling issues in ecodesign constrained by expected future scarcity of metals. Three criteria were given for this specific application of ecodesign: increased recyclability, decreased dependency on material supply risk, and reduced environmental impacts.

In the first case of Cu substitution, the use of Al or carbon fibres helps reducing the material dependency of future EV manufacturers on a single raw material source. While easy to recycle as such, the further use of Al depends heavily on the alloying metals used in the specific application. Alloying with manganese, for instance, severely limits the re-use potential of Al [5]. Recycling of carbon fibres is challenging and the performance of recycled carbon fibres not adequate for the most demanding applications [5]. Finally, the use of Al or carbon fibres in EVs could lead to reduced vehicle weight, thus reducing fuel use and improving the environmental performance of driving. This means that two of our three criteria on ecodesign are well met by the Cu substitution case. To make even a stronger case, EV designers should judge also the after-life of the materials used in electric motors.

The second case of spray coating shows a better recyclability as it reduces the use of Ni alloyed with nyobium and tantalum. Currently less than one per cent of tantalum is recycled, while niobium enjoys a recycling rate of more than 50 per cent on a global average [5]. While the alloying elements can, in principle, be recovered from the slag phase after the electric arc furnace (EAF) process in a steel mill, their concentrations are not likely to be high enough for economically feasible recovery. As for material supply risk, both niobium and tantalum belong to the critical materials category in recent EU categorization. Thus, avoiding these elements in boiler design would make new power plant production less vulnerable for market fluctuation.

Finally, the environmental performance of spray coating is likely to have similar or reduced impact on the environment in comparison to the original design.

REFERENCES

1 Hashimoto, S., Fischer-Kowalski, M., Suh, S., Bai, X. 2012. Greening Growing Giants. A Major Challenge of Our Planet. Journal of Industrial Ecology 26 (4): 459-466.
2 Ni Study Group: World Ni Statistics 2012; International Cu Study Group: Statistical Yearbook 2011
3 Enterprise and Industry Directorate General, 2010: Critical raw materials for the EU
4 McDonough, W., Braungart M. 1998. The NEXT Industrial Revolution. Atlantic Magazine, October 1998; Betts, K. 1998. Preventing Pollution by Design. Environmental Science & Technology 32(13): 318A-320A
5 Reck, B.K., Graedel, T.E. 2012. Challenges in Metal Recycling. Science 337: 690-695
6 Cuinfo, Recycling of Cu. http://www.Cuinfo.co.uk/environment/recycling.shtml
7 Diwekar, U.M. 2003. Greener by Design. Environmental Science & Technology 37(23): 5432-5444; Gaustad, G. et.al. 2010. Design for Recycling – Evaluation and Alloy Modification. Journal of Industrial Ecology 14(2): 286-308.
8 Sperling, D. and Gordon, D. 2010. Two Billion Cars: Driving toward Sustainability. Oxford University Press.
9 http://www.Cu.org/about/pressreleases/2010/pr2010_March_5.html
10 Buller, U. 2011. Alternativen bei drohender Rohstoffknappheit (Alternatives to the impending raw material scarcity). Paper presented at 22. Automobilforum Stuttgart – 24 May 2011. Available online: http://www.automobil-produktion.de/uploads/2011/05/1645-Buller.pdf
11 Kampman, B., van Essen, H. et.al. 2011. Impact analysis on market uptake scenarios and policy implications. Impacts of Electric Vehicles – Deliverable 5. Delft, CE Delft, April 2011. Available online: http://ec.europa.eu/clima/policies/transport/vehicles/docs/d5_en.pdf
12 Miller T.J.E. 2002. Optimal Design of Switched Reluctance Motors. IEEE Trans. Ind. Electron., 49(1). pp. 15-27.
13 Halmeaho T. 2012. Magnetic bearing as switched reluctance motor. Master's thesis. 86+12 p.
14 Han, P-W., Chun, Y-D, Choil, J-H, Kiml, M-J., Kool D-H. and Lee, J. The study to Substitute Aluminum for Cu as a winding material in Induction Machine. lIndustry Applications Research Division, Korea Electrotechnology Research Institute, Changwon, Korea, E-mail: pwhan@keri.re.kr. http://ieeexplore.ieee.org/stamp/stamp.jsp?arnumber=05352008
15 Jarosz, P., Schauerman, C., Alvarenga, J., Moses, B., Mastrangelo, T., Raffaelle, R., Ridgley, R and Landi, B. 2011. Carbon nanotube wires and cables: Near-term applications and future perspectives Received 13th July 2011, Accepted 8th September 2011 DOI: 10.1039/c1nr10814j
16 Suhonen, T. and Varis, T. 2012. Thermal Spray Coating. Patent WO2012131164 (A1).

Mater. Res. Soc. Symp. Proc. Vol. 1492 © 2013 Materials Research Society
DOI: 10.1557/opl.2013.319

Environmental Impacts of Distributed Manufacturing from 3-D Printing of Polymer Components and Products

Megan Kreiger[1] and Joshua M. Pearce[1,2]
[1]Department of Materials Science & Engineering, Michigan Technological University, Houghton, MI, United States.
[2] Department of Electrical & Computer Engineering, Michigan Technological University, Houghton, MI, United States.

ABSTRACT

Although additive layer manufacturing is well established for rapid prototyping the low throughput and historic costs have prevented mass-scale adoption. The recent development of the RepRap, an open source self-replicating rapid prototyper, has made low-cost 3-D printers readily available to the public at reasonable prices (<$1,000). The RepRap (Prusa Mendell variant) currently prints 3-D objects in a 200x200x140 square millimeters build envelope from acrylonitrile butadiene styrene (ABS) and polylactic acid (PLA). ABS and PLA are both thermoplastics that can be injection-molded, each with their own benefits, as ABS is rigid and durable, while PLA is plant-based and can be recycled and composted. The melting temperature of ABS and PLA enable use in low-cost 3-D printers, as these temperature are low enough to use in melt extrusion in the home, while high enough for prints to retain their shape at average use temperatures. Using 3-D printers to manufacture provides the ability to both change the fill composition by printing voids and fabricate shapes that are impossible to make using tradition methods like injection molding. This allows more complicated shapes to be created while using less material, which could reduce environmental impact.

As the open source 3-D printers continue to evolve and improve in both cost and performance, the potential for economically-viable distributed manufacturing of products increases. Thus, products and components could be customized and printed on-site by individual consumers as needed, reversing the historical trend towards centrally mass-manufactured and shipped products. Distributed manufacturing reduces embodied transportation energy from the distribution of conventional centralized manufacturing, but questions remain concerning the potential for increases in the overall embodied energy of the manufacturing due to reduction in scale. In order to quantify the environmental impact of distributed manufacturing using 3-D printers, a life cycle analysis was performed on a plastic juicer. The energy consumed and emissions produced from conventional large-scale production overseas are compared to experimental measurements on a RepRap producing identical products with ABS and PLA. The results of this LCA are discussed in relation to the environmental impact of distributed manufacturing with 3-D printers and polymer selection for 3-D printing to reduce this impact. The results of this study show that distributed manufacturing uses less energy than conventional manufacturing due to the RepRap's unique ability to reduce fill composition. Distributed manufacturing also has less emissions than conventional manufacturing when using PLA and when using ABS with solar photovoltaic power. The results of this study indicate that open-source additive layer distributed manufacturing is both technically viable and beneficial from an ecological perspective.

INTRODUCTION

Since the dawn of the industrial age, manufacturing has been about internally maximizing profit. This search for maximum profit created a general trend in towards large-scale manufacturing in low-labor cost countries, especially for inexpensive plastic products [1-3]. Recently, several open-source (OS) models of commercial rapid prototypers have been developed [4], which offer an alternative model of low-cost production. Historically, commercial rapid prototypers or 3-D printers use additive layer manufacturing (AM) to accurately fabricate custom or generic products out of various materials including plastic, metal or ceramics. The technological development of 3-D printers has been substantial, which has benefited many industries [5-9]. However, the costs of 3-D printers have historically been too expensive to be feasible for distributed or home-based manufacturing. With the introduction of the RepRap, an OS 3-D printer that can be built for under $500, this has changed. The RepRap, has opened the door of AM to a wide range of potential users due to cost and simplicity [10] and made distributed small-scale production technically feasible [11].

The RepRap has a 200 mm x 200 mm x 140mm (height) build envelope and can currently print using various plastics. The plastics typically used are acrylonitrile butadiene styrene (ABS) and polylactic acid (PLA). ABS and PLA are both thermoplastics that can be injection-molded, each with their own benefits, as ABS is rigid and durable, while PLA is plant-based and can be recycled and composted. The melting temperatures of PLA and ABS allow for safe extrusion, while being high enough to ensure shape retention. Distributed recycling is also being developed to recycle post-consumer products into filament for a 3-D printer, which could further reduce cost and resources required for distributed manufacturing [12].

The use of 3-D printers allows for previously impossible shapes under conventional manufacturing methods (e.g. injection molding) along with the ability to manipulate the inside of an object in multiple ways, such as, fill composition or adding internal parts. The ability to manipulate shapes internally during production has the potential to reduce additional machining during processing. Holes, voids, and other needs within an object that were impossible using methods similar to injection molding previously had to be done using tools, such as, drill presses. These steps are now able to be created during the design step and automatically produced using the RepRap. The ability to change fill composition allows more complicated shapes to be produced with structural integrity with the use of less material. This property combined with the reduction in embodied energy of transportation made available by distributed manufacturing allow for the possibility that it could be less energy and emission intensive than conventional manufacturing. However, questions remain about the environmental benefits of distributed manufacturing due the potential for increases in the overall embodied energy of the manufacturing due to reduction in scale.

This preliminary study explores these questions by probing the technical potential of using a distributed network of RepRaps to produce goods. A life cycle analysis (LCA) of energy consumption and emissions is performed for distributed manufacturing using RepRaps and compared to conventional manufacturing overseas. These results are discussed to draw conclusions about the viability of distributed manufacturing.

METHODS

The RepRap (Prusa Mendell variant) was used to print all product/product components using both ABS and PLA (Fig. 1) using 15% fill in order to reduce the amount of plastic used to produce it while maintaining the necessary functional mechanical strength. The ABS extruder temperature was 230°C and bed temperature was 110°C. PLA was printed using an extruder temperature of 185°C, with a first layer bed temperature of 63°C to ensure adhesion followed by print bed temperature of 60°C. Energy measurements were taken using a multimeter (±0.005 kWh) during initial heating and while printing each individual object.

The product evaluated was a juicer (Fig. 2). The juicer is used to juice oranges, grapefruits, lemons, limes and other citrus fruit. It is fixed upon a post-consumer glass jar to collect the juice. Typical manufacturing is limited on internal manipulation of a product, with the use of the RepRap, this is no longer a barrier. The fill can be altered by pattern (rectilinear, line, concentric, honeycomb, etc), angle, percentage, along with adding solid fill layers when necessary. In this study, all products were printed with the fill at 45 degrees using the rectilinear pattern and the setting of adding solid layers when necessary. The life-cycle analysis (LCA) software SimaPro 7.2 was used to get the cumulative energy demand (CED 1.07) and greenhouse gas emissions (IPCC GWP 2007 100a) for each product using the database EcoInvent v2.0. A "cradle-to-gate" analysis was done, with the gate located in the United States.

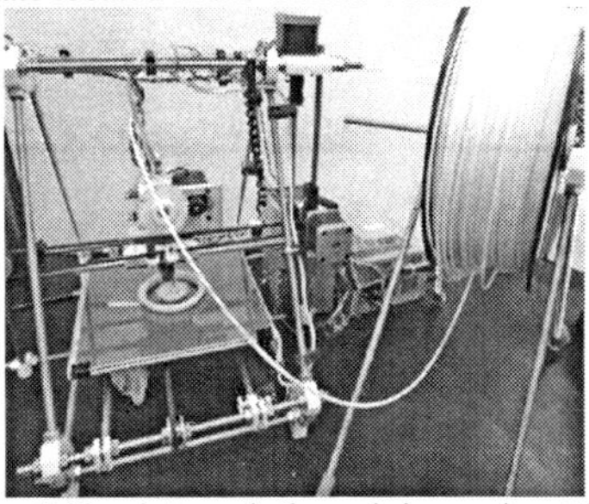

Fig. 1: Prusa Mendell RepRap 3-D printer printing a juicer using ABS.

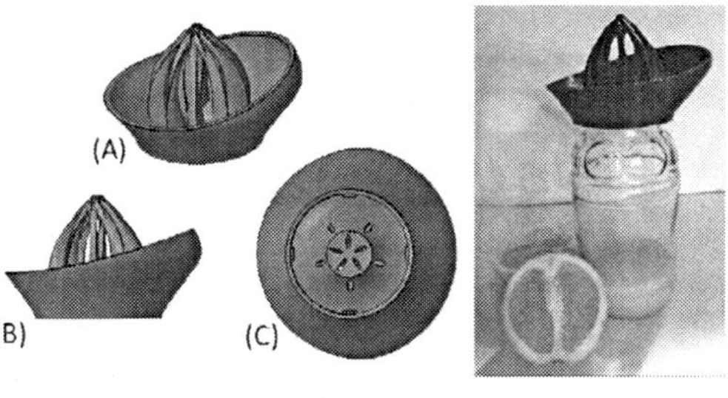

Fig. 2: Top-side (A), side (B) and bottom (C) views of CAD for a citrus juicer in photograph (STL available: http://www.thingiverse.com/thing:21100 based on original design by S. Bernier).

Distributed manufacturing was quantified using the electricity consumption of the RepRap and the material inputs by weight for ABS or PLA, where (Electricity, Production Mix, US) or (Electricity, PV, at 3kWp, a-Si panel, Switzerland) was used as an input, in addition to

(Polylactide, NatureWorks Nebraska, US) and (Acrylonitrile-butadiene-styrene copolymer, Switzerland).

Conventional manufacturing was input into SimaPro assuming a 100% fill and was input based on the weight from experimental results. The inputs used for conventional were material inputs PLA or ABS, as done for distributed, in addition to injection molding (Injection molding, Switzerland) and shipping 9,213 km using (Transoceanic freight ship, international) from Shanghai, China to Seattle, WA [13]. The injection molding input states that the finished product is 99.3% of its input, this was considered in the analysis for both the material input and the injection mold.

Processing for each of these cases should be assumed to underestimate the total cumulative energy demand and emissions, as additional processing may be required for consumer use (i.e. sanding, finishing, etc.). Overseas shipping distance is an underestimate due to taking a straight-line trip across the ocean. Shipping over land, infrastructure, molds, packaging and waste were not included in this analysis thus underestimating the embodied energy of traditional manufacturing and are left for future work. The materials PLA and ABS were used as an example for each product, but may not be the ideal materials for these products or may require additional coatings to make them food-grade or child-safe.

RESULTS AND DISCUSSION

Measured experimental values from the RepRap are 0.31 kWh for the PLA and 0.52 kWh for the ABS juicers respectively. These values were input into SimaPro and compared to conventional methods as described for cumulative energy demand (CED) (Fig. 3) and greenhouse gas emissions (Fig. 4). Conventional PLA and ABS juicers at 100% fill had CED of 11.58 and 13.71 MJ, respectively. Distributed PLA and ABS juicers at 15% fill were found to have a CED of 8.66 and 12.96 MJ using conventional electricity.

The environmental impacts were minimized using a solar photovoltaic array (PV) to provide electricity following recommendations by [11] that would allow for distributed manufacturing of products with RepRap in most locations in the world. It has been well established that PV technology is a sustainable source of energy that significantly reduces environmental impact of electricity use and is amenable to distributed generation [14]. Although there are no commercial PV-powered RepRap 3-D printers, proof of concepts already exist and the open-source development community that supports the RepRap has been experimenting with variants [15]. These variants would enable distributed manufacturing even in remote communities without access to the conventional grid.

The addition of a PV system reduced the values of the CED to 6.32 and 9.03 MJ, respectively. Conventional manufacturing using PLA and ABS had the emissions of 0.43 and 0.62 kg CO_2 eq, while the distributed case with the use of conventional centralized grid electricity had 0.39 and 0.68 kg CO_2 eq and when powered with PV: 0.18 and 0.32 kg CO_2 eq. The energy is minimized using distributed manufacturing for the juicer and is made possible by using a smaller fill percentage. This not only reduces material use in the product itself but also the environmental impact of the processing and embodied energy use in the raw material extraction and transportation. The use of PV to power the RepRap minimizes both the emissions and the energy use. The emissions are lower for the distributed manufacturing systems for all cases except the ABS juicer without PV. This is due to the relatively large amount of energy needed to keep the heated build platform at operating temperature for the ABS. Future work is necessary to reduce the energy needed for the build platform. This can be done by using chemical means to enable better adhesion, using zoned heating so only the part of the bed that is needed for the part is heated, better insulating the bed, or using a controlled environmental chamber to insulate the entire RepRap from cold ambient temperatures.

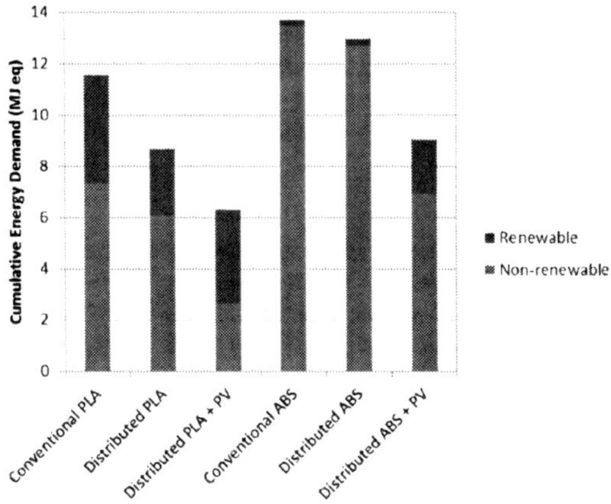

Fig. 3: CED of the juicer showing conventional PLA and ABS at 100% fill, distributed PLA and ABS at 15% fill, along with the effect of PV electricity. Renewable consists of renewable bio-mass, wind, solar, and water energy sources. Non-renewable consists of non-renewable energy sources fossil fuels, nuclear, and bio-mass.

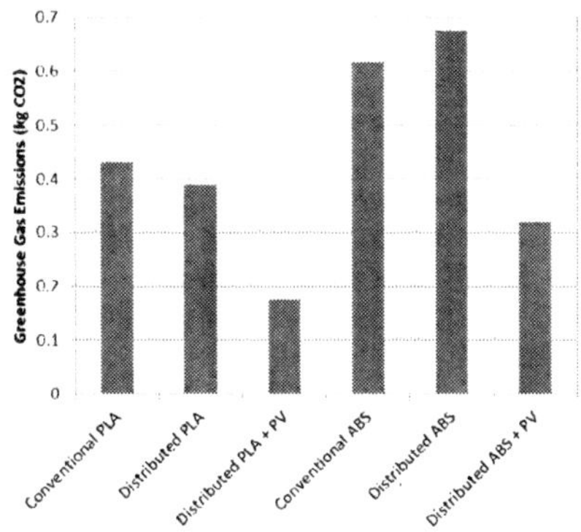

Fig. 4: Greenhouse gas emissions in kg CO_2 eq (GWP 100a) for the juicer for conventional PLA and ABS 100% fill, distributed PLA and ABS at 15% fill, along with the effect of PV electricity.

The results of this study show that distributed manufacturing requires less cumulative energy than conventional manufacturing when products are made from PLA and ABS. If more products are printed simultaneously on the heating bed, it may be possible to reduce the energy to print even further due to the initial heating energy being dispersed among more individual products. When distributed manufacturing is used in conjunction with a PV system, the cumulative energy is further decreased. Distributed manufacturing creates fewer emissions than conventional when PLA is used and in cases using PV electricity with ABS. The emissions are highly dependent on the grid emission intensity. The addition of a PV system reduces emissions in every case and future work is needed in this area of distributed generation of electricity for distributed manufacturing. This was a preliminary study on a single product and significant future work is necessary to quantify the LCAs of distributed vs. conventional manufacturing of other products. In addition, it was clear from this preliminary study that the fill percentage plays a critical role in the environmental impact and should be systematically explored.

CONCLUSIONS

The results of this LCA study indicate that the environmental impact of distributed manufacturing with 3-D printers of polymer objects is less than conventional when using PLA. This indicates that distributed manufacturing is technically viable and environmentally beneficial because of reduced energy consumption and greenhouse gas emissions. These positive environmental results for distributed manufacturing are expanded to ABS when PV is used as a source of electricity. The preliminary results indicate that the ability of RepRaps and similar 3-D printers to vary fill percentage has the potential to significantly diminish environmental impact on many products. It can be concluded from the results of this study that open-source additive layer distributed manufacturing is both viable and beneficial from an ecological perspective.

ACKNOWLEDGMENTS

The authors would like to acknowledge helpful discussions with D.R. Shonnard, B. Tymrak and G. Anzalone. This research was supported by Sustainable Futures Institute.

REFERENCES

1. I. Kravis, R. Lipsey. *J. of Int. Econ.* **12**(3), 201 (1982).
2. S. Thurm. *Wall Street Journal.* April (2012).
3. K. Ruamsook, D. Russell, E. Thomchick. *J. of Supply Chain Management.* **43**,16 (2007).
4. R. Jones, P. Haufe, E. Sells. *Robotica.* **29**, 1 (2011).
5. S. Upcraft, R. Fletcher. *Assembly Automation.* **23**, 4 (2003).
6. I. Gibson, D.W. Rosen, B. Stucker. *Physics Procedia.* **5** (2010).
7. V. Petrovic, J.V.H. Gonzalez, O.J. Ferrando, J.D. Gordillo, J.R.B. Puchades, L.P. Griñan. *Int. J. of Production Research.* **49**, 4 (2010).
8. A. Gebhardt, F. Schmidt, J. Hötter, W. Sokalla, P. Sokalla. *Physics Procedia.* **5**, 2 (2010).
9. N.B. Crane, J. Tuckerman, G.N. Nielson. *Rapid Prototyping Journal.* **17**, 3 (2011).
10. RepRap. WWW document, (reprap.org/wiki/Main_Page)
11. J. M Pearce, C. Morris Blair, K. J. Laciak, R. Andrews, A. Nosrat, I. Zelenika-Zovko, *J. of Sust. Dev.* **3**(4), 17 (2010).
12. C. Baechler, M. DeVuono, J.M. Pearce. *Rapid Prototyping.* **19**(2) (2013).
13. City distance calculator. WWW document (http://www.geobytes.com/citydistance.htm)
14. J. Pearce. *Futures* **34**(7), 663 (2002).
15. D. King and J.M. Pearce. WWW document (http://www.appropedia.org/Mobile_solar_powered_3D_printer)

Mater. Res. Soc. Symp. Proc. Vol. 1492 © 2013 Materials Research Society
DOI: 10.1557/opl.2013.258

Distributed Recycling of Post-Consumer Plastic Waste in Rural Areas

M. Kreiger,[1] G. C. Anzalone[2], M. L. Mulder[1], A. Glover[1], and J. M Pearce[1,3,*]
[1]Department of Materials Science & Engineering, Michigan Technological University, Houghton, MI, United States.
[2] Department of Civil & Environmental Engineering, Michigan Technological University, Houghton, MI, United States.
[3] Department of Electrical & Computer Engineering, Michigan Technological University, Houghton, MI, United States.
[*] corresponding author:pearce@mtu.edu

ABSTRACT

Although the environmental benefits of recycling plastics are well established and most geographic locations within the U.S. offer some plastic recycling, recycling rates are often low. Low recycling rates are often observed in conventional centralized recycling plants due to the challenge of collection and transportation for high-volume low-weight polymers. The recycling rates decline further when low population density, rural and relatively isolated communities are investigated because of the distance to recycling centers makes recycling difficult and both economically and energetically inefficient. The recent development of a class of open source hardware tools (e.g. RecycleBots) able to convert post-consumer plastic waste to polymer filament for 3-D printing offer a means to increase recycling rates by enabling distributed recycling. In addition, to reducing the amount of plastic disposed of in landfills, distributed recycling may also provide low-income families a means to supplement their income with domestic production of small plastic goods. This study investigates the environmental impacts of polymer recycling. A life-cycle analysis (LCA) for centralized plastic recycling is compared to the implementation of distributed recycling in rural areas. Environmental impact of both recycling scenarios is quantified in terms of energy use per unit mass of recycled plastic. A sensitivity analysis is used to determine the environmental impacts of both systems as a function of distance to recycling centers. The results of this LCA study indicate that distributed recycling of HDPE for rural regions is energetically favorable to either using virgin resin or conventional recycling processes. This study indicates that the technical progress in solar photovoltaic devices, open-source 3-D printing and polymer filament extrusion have made distributed polymer recycling and upcycling technically viable.

INTRODUCTION

Global production of plastic increased by 500% over the last 30 years and it is expected to continue to grow to 850 million tons/year by 2050 [1-3]. Plastic use results in a substantial environmental burden due to both land and water pollution as plastics take 10 to 450 years to decompose in landfills [4-6]. Plastic processing, use, and disposal also comprise a significant source of energy consumption with the concomitant emissions and additional pollution [7-10]. Rural areas in the U.S. generally have limited recycling opportunities, which makes recycling difficult and burdensome to rural residents, which make up ~ 19% of the population (60m) [11].

Two recent open-source hardware technological developments, 3-D printers and RecycleBots, offer a new approach to polymer recycling (or upcycling), encompassing the potential for distributed processing to high-value added products, which reverses the historical trend towards centralized manufacturing and recycling facilities. Application of these technologies could induce job creation and allow for residents of rural areas to supplement their income by creating and customizing goods that were historically not available locally.

Highly accurate commercial 3-D printers are useful production and design tools. The development of additive manufacturing for rapid prototyping and 3-D printing in a number of technologies has been substantial [12-16]. Recently, following the open source (OS) model, the RepRap has been developed which can be built for ~$500, greatly expanding the potential user base of 3-D printers. These machines could feasibly be used for small-scale manufacturing in isolated regions or as an enabling tool for green manufacturing [17]. Commercial extrusion of plastic utilizes a screw to move material through a heated barrel where it is compressed, melted, mixed and forced through a die [18]. A small-scale plastic extruder using these principles to turn waste plastic into a growth medium for plants [19], has been modified to create a new, semi-automated open source "RecycleBot" to prepare RepRap feedstock from household plastic waste such as milk jugs and laundry detergent containers [20].

The primary expense of operating a 3-D printer is the filament or "3-D ink" and thus the operating costs and environmental impact of the RepRap can be further reduced using waste plastics as feedstock and may provide an incentive for distributed, in-house recycling of plastic waste [17]. Baechler et al., have demonstrated acceptable 3-D filament production from a RecycleBot using high density polyethylene (HDPE) [20]. Several other open-source filament fabricators have also been recently developed (e.g. Filabot [21], Lyman Filament Extruder[22]), which could be utilized as waste plastic RecycleBots. HDPE, recycled plastic number "2", is used primarily for packaging, pipes, and plastic lumber. HDPE poses known challenges for recycling, but with the addition of a stabilizing system the recycled output can have similar properties to virgin resin [23].

This study explores the technical potential of using a distributed network of RecycleBots to process HDPE waste into 3-D printing feedstock. A life cycle analysis (LCA) of energy consumption is performed for this distributed approach and compared to the conventional recycling in rural areas. In this radical distributed approach the recycling process occurs at the site of consumption, with the plastic processed by the consumer. These results are discussed to draw conclusions about the viability of distributed recycling.

METHODS

The distributed recycling process includes: cleaning, cutting, and shredding plastic bottles and other plastic recyclables of the same plastic type (Fig. 1a). These processes are assumed to be completed by the consumer on site and are not included in the analysis. The shreds of plastic are put into a waste plastic extruder (RecycleBot: Fig. 1b) described in detail by Baechler et al. [20]. The RecycleBot melts down the plastic and creates a 3mm filament that can be used in a RepRap or other type of 3-D printer (Fig. 1c). The energy consumption for the RecycleBot was measured during extrusion with a multimeter (±0.005 kWh). The stages included in this study were auger drive and heating (shredding was previously shown to have negligible energy consumption using less than 3% of the energy of the other stages and could be accomplished manually [20]).

The LCA software SimaPro 7.2 was used to get the cumulative energy demand (CED 1.07) for each case along with the database EcoInvent v2.0. For distributed recycling based on the electricity usage of the RecycleBot, (Electricity, Production Mix, US) was used as an input to make the results of this study as generalizable as possible. The cut-off method was used in each case, where the initial virgin resin production is not considered in the recycled energy demand. This described method using SimaPro was used for the experimental RecycleBot value, in order to be able to do a direct comparison between each method of recycling HDPE.

For the rural area, the small, geographically isolated town of Copper Harbor, MI was used. Located at the tip of the Keweenaw Peninsula, Copper Harbor is a 48 mile drive from the nearest recycling collection center in Houghton, MI and there is no curbside pickup. Plastic waste is transported by semi-truck from Houghton to the processing center in Green Bay, WI, 212 miles away. The average household generates 16.9 pounds of mixed recyclables per week of which 5.2 pounds or 31% is HDPE [24]. Two scenarios are considered: recycling (1) biweekly or (2) monthly where transportation is provided by passenger vehicles with a mixed recyclable load. SimaPro was used to determine the cumulative energy demand using (Operation, passenger car, petrol, fleet average 2010, Switzerland). The input used for the Houghton to Green Bay trip was (Operation, lorry 3.5-20t, full, fleet average, Switzerland) the return trip was considered using a similar input for (empty). The Swiss data has been previously shown to be reliable and can be expected to representative of U.S. energy uses. All of these values were considered for 1 kg of recycled HDPE, to account for the 8% waste in processing [24].

Fig. 1: Distributed HDPE Recycling. a) Cutting and shredding HDPE milk jug, b) Extruding 3-D printing filament in RecycleBot (improved performance was observed wrapping the front and back of the extruder and the insulation with Kapton tape, not shown here for clarity), c) RepRap 3-D printer self-replicating.

An LCA study of conventional HDPE recycling was previously completed [24] using confidential information from recycling companies and was used here to quantify the impact and energy demand for conventional centralized recycling. The energy from the transport of plastics (collection) was added to the additional energy demand required to process the plastic of 7.513 MJ per kg HDPE recycled, using the cut-off weight based method excluding collection [24], to give an estimate of the total cumulative energy demand.

RESULTS AND DISCUSSION

Table 1 summarizes the energy demand for distributed recycling of HDPE using the distributed RecycleBot, HDPE virgin resin, and the conventional method of recycling in rural areas under the conditions of biweekly or monthly trips to the collection center.

In the distributed recycling case the RecycleBot required 0.06 kWh for initial heating, and 0.0036 kWh/m of filament. The total energy used for filament production was 0.694 kWh (2.5 MJ) per kg of recycled HDPE filament. 2.5 MJ per kg recycled HDPE was input into SimaPro 7.2 using cumulative energy demand 1.07 and evaluated with the only input being the electricity production mix of the U.S., from the EcoInvent 2.0 database, which gives the cumulative energy demand of 8.74 MJ shown in Table 1.

Table 1: Energy Demand and Reduction for Various Recycling Cases

Case	Energy Demand (MJ/kg HDPE)	Percent Reduction (Δ%) for Distributed Recycling [c]
Distributed Recycling: Insulated RecycleBot	8.74	--
Virgin Resin[a]	79.7	89
Centralized Recycling[b] – Rural: Copper Harbor (monthly)	28.4	69
Centralized Recycling[b] – Rural: Copper Harbor (bi-weekly)	48.9	82

Notes: a. [26], b. Estimate based on [24],
c. Percent reduction = (Central-Distributed)/Central*100

Energy demand values in the RecycleBot cases are higher than the experimental measurements as they include the additional energy to produce the electricity used. It should be pointed out that, the virgin resin embodied energy of 79.67 MJ/kg HDPE [25], underestimates the energy required, as additional energy is needed to turn the resin into filament. Even with this underestimate, distributed recycling reduces the energy demand of HDPE over the virgin resin by a substantial 89%.

When comparing the conventional rural recycling to distributed recycling, distributed recycling can reduce the energy demand by at least 69-82% using conventional electricity in rural areas such as the case study. Monthly and biweekly trips for recycling result in an energy demand of at least 28.4 MJ and 48.9 MJ per kg of recycled HDPE. For rural locations, the energy values for transport can only be reduced by transporting more materials per trip, hence the potential for taking monthly trips instead of biweekly trips. If conventional recycling is used, the best option is to transport as many recyclables as possible and minimize the frequency of these trips.

A sensitivity analysis was done based on the miles transported for the rural case as seen in Fig. 2. It was found that for rural locations over one mile from the nearest collection center that will need to make trips biweekly, it requires less energy to use distributed recycling powered conventionally. For rural locations over two miles from the nearest collection center, less energy is required for distributed recycling than conventional, even when only making monthly trips.

The cumulative energy of the distributed case could be further reduced using a low-embodied energy distributed generation technology such as solar photovoltaic (PV) systems [25]. Conventional cases could be similarly improved, however, the distributed nature of PV couples favorably with the distributed recycling technology. As the costs have decreased for distributed

PV [27], a PV-powered distributed polymer recycling process becomes potentially economically as well as energetically viable. This technical progress coupled to new peer-to-peer financing provides the potential for rural communities to become more autonomous while improving local economies [28]. Work is underway now to merge these technologies at several institutions.

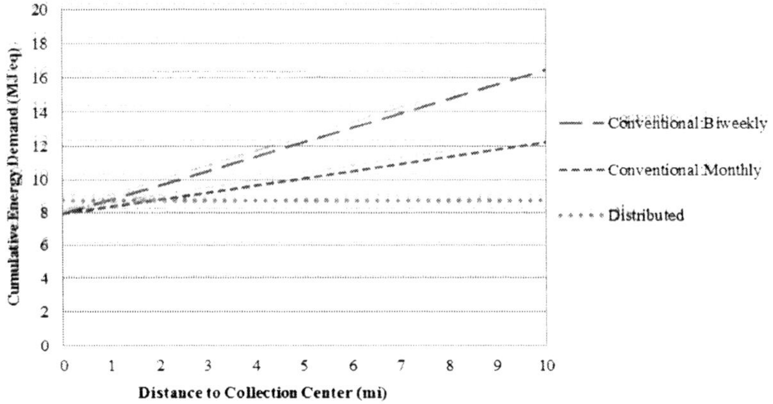

Fig. 2: Energy demand vs distance from recycling center (biweekly and monthly trips) compared to distributed recycling using the RecycleBot (Distributed).

The recycled filament could be used to print any of the ~50,000 open-source designs in the Thingiverse database on in-home 3-D printers or it could be sold on the open market. Future work on the energy demand of distributed recycling of other plastics that can be used in 3-D printers needs to be considered. Also, future work on tuning printing with HDPE needs to be accomplished, potentially necessitating the design of an open-source environmental chamber.

CONCLUSIONS

The results of this LCA study indicate that distributed recycling of HDPE for rural regions is energetically favorable to either using virgin resin or conventional recycling processes. In the case study explored, embodied energy savings of 69 to 82% were found for RecycleBot technology over centralized recycling. These results provide further support for minimizing transportation distances for recycling to minimize environmental impact. This study indicates that the technical progress in solar photovoltaic devices, open-source 3-D printing and polymer filament extrusion have made distributed polymer recycling and upcycling technically viable and potentially economically advantageous for rural residents in both the US, but also the ~ 3 billion people globally living in rural areas.

ACKNOWLEDGMENTS

The authors would like to acknowledge technical support from A. Vora. This research was supported by Sustainable Futures Institute and a McArthur Research Internship.

REFERENCES

1. L.C.-M. Lebreton, S.D. Greer, J.C. Borrero. Marine Pollution Bulletin. **54**, 3 (2012).
2. L. Shen, J. Haufe, M.K. Patel. Product Overview and Market Projection of Emerging Bio-based Plastics PRO-BIP 2009, Utrecht University Final Report, (2009).
3. A. Lotfi, Polymer Recycling. WWW document, (http://www.lotfi.net/recycle/plastic.html).
4. J.F. Rees. J. Chem. Technol. Biotechnol. **30**, 1 (1980).
5. J.G.B. Derraik. Marine Pollution Bulletin. **44** (2002).
6. U.S. National Park Service. Time it takes for garbage to decompose in the environment. WWW document, (http://des.nh.gov/organization/divisions/water/wmb/coastal/trash/documents/marine_debris.pdf).
7. U. Arena, M.L. Mastellone, F. Perugini. Int. J. of Life Cycle Assessment. **8**, 2 (2003).
8. F. Perugini, M.L. Mastellone, U. Arena. Environ. Prog. **24**, 2 (2005).
9. P.M. Subramanian. Resources, Conservation and Recycling. **28**, 3-4 (2000).
10. A. Björklund, G. Finnveden. Resources, Conservation and Recycling. 44, 4 (2005)
11. U. S. Census Bureau, 2010 Census Urban and Rural Classification and Urban Area Criteria. WWW doc. (http://www.census.gov/geo/www/ua/2010urbanruralclass.html).
12. S. Upcraft, R. Fletcher. Assembly Automation. **23**, 4 (2003).
13. I. Gibson, D.W. Rosen, B. Stucker. Physics Procedia. **5** (2010).
14. V. Petrovic, J.V.H. Gonzalez, O.J. Ferrando, J.D. Gordillo, J.R.B. Puchades, L.P. Griñan. Int. J. of Production Research. **49**, 4 (2010).
15. A. Gebhardt, F. Schmidt, J. Hötter, W. Sokalla, P. Sokalla. Physics Procedia. **5**, 2 (2010).
16. N.B. Crane, J. Tuckerman, G.N. Nielson. Rapid Prototyping Journal. **17**, 3 (2011).
17. J. M Pearce, C. Morris Blair, K. J. Laciak, R. Andrews, A. Nosrat, I. Zelenika-Zovko, J. of Sust. Dev. **3**(4), 17 (2010).
18. D. Rosato. Plastics Processing Data Handbook. 2nd ed.(Springer-Verlag, 1997).
19. R. Torcellini. Plastic extruder for growing media. WWW document, (http://web4deb.blogspot.com/2010/12/plastic-extruder-for-growing-media.html).
20. C. Baechler, M. DeVuono, J.M. Pearce. Distributed recycling of waste polymer into reprap feedstock. Rapid Prototyping Journal, **19**(2) (in press) March 2013.
21. Filabot Personal Filament Maker. WWW document, (http://filabot.com).
22. Lyman Filament Extruder. WWW document, (http://www.thingiverse.com/thing:30642).
23. M. Loultcheva, M. Proietto, N. Jilov, F.P. La Mantia. Polymer Degradation and Stability. **57**, 1 (1997).
24. Franklin Associates, Life cycle inventory of 100% postconsumer HDPE and PET recycled resin from postconsumer containers and packaging, (2011).
25. G. Hammond, C. Jones. Inventory of carbon & energy (ICE) Version 1.6a,Bath U.(2008).
26. J. Pearce, A. Lau, Net Energy Analysis For Sustainable Energy Production From Silicon Based Solar Cells. *Proc. of A. Soc. of Mech. Eng. Solar*, Cambell-Howe (Ed.), (2002).
27. K. Branker, M. J.M. Pathak, J. M. Pearce, A Review of Solar Photovoltaic Levelized Cost of Electricity", *Renewable & Sustainable Energy Reviews* **15**, 4470 (2011).
28. K. Branker, E. Shackles, J. M. Pearce, Peer-to-Peer Financing Mechanisms to Accelerate Renewable Energy Deployment. J. of Sust. Finance & Investment **1**(2), 138 (2011).

Smart Materials

Mater. Res. Soc. Symp. Proc. Vol. 1492 © 2013 Materials Research Society
DOI: 10.1557/opl.2013.155

Progress in Chromogenic Materials and Devices: New Data on Electrochromics and Thermochromics

C. G. Granqvist, S.-Y. Li, İ. Bayrak Pehlivan, and G. A. Niklasson
Department of Engineering Sciences, The Ångström Laboratory, Uppsala University,
P. O. Box 534, SE-75121 Uppsala, Sweden

ABSTRACT

Electrochromic (EC) and thermochromic (TC) materials are of much interest for "smart" windows which combine energy efficiency with the provision of indoor comfort. This paper summarizes results from several recent studies related to nanoparticles of transparent and electrically conducting ITO (*i.e.*, In_2O_3:Sn) and of thermochromic VO_2. Specifically, we consider (*i*) the use of ITO nanoparticles in polaronic EC devices in order to suppress near-infrared solar transmittance, (*ii*) performance limits for plasmonic EC devices embodying ITO nanoparticles, and (*iii*) ITO-VO_2-based nanocomposites with joint low thermal emittance and TC properties, and with Mg-doping of the VO_2 as a means for boosting the luminous transmittance. Both experimental and theoretical results are presented.

INTRODUCTION

Windows are necessary elements in buildings where people live and work, and they give visual indoors-outdoors contact along with day-lighting. However, the windows usually let in or out too much energy, which must be balanced by energy guzzling space heating or space cooling. Multiple glazing is necessary to diminish heat losses and provide energy efficiency, and a transparent low-emittance coating on one of the surfaces enclosing a gas-filled gap between two neighboring panes can bring down the heat transfer to roughly half of what it would be in the absence of the coating [1]. A further significant improvement of the window can be achieved if it also is "smart", meaning that it admits solar energy into a room when there is a heating demand and prevents the solar energy inflow—but keeps unimpeded indoors-outdoors contact—when there is a cooling demand. The "smartness" can be accomplished by chromogenic fenestration [2] involving electrochromic (EC) thin films, allowing electrically controlled inflow of visible light and solar energy, or a thermochromic (TC) thin film which limits the solar energy inflow above a certain "comfort temperature".

This paper summarizes some recent studies of ours which involve nanoparticles of transparent and electrically conducting tin doped indium oxide (In_2O_3:Sn, known as ITO) and of thermochromic VO_2, sometimes with varying degrees of Mg doping. The first study regards a standard ("polaronic") EC device incorporating a transparent polymer electrolyte and demonstrates experimentally and theoretically that the near-infrared (NIR) part of the solar irradiation—with about half of the solar energy—can be absorbed very efficiently by ITO nanoparticles dispersed in the electrolyte, and that this functionality does not interfere much with an EC modulation of visible light [3]. The second study considers a new type ("plasmonic") EC device, which embodies ITO nanoparticles, and presents computed performance limits on the optical modulation [4]. Thirdly we regard nanocomposites of ITO and thermochromic VO_2 and demonstrate that a temperature-dependent modulation of the near-infrared transmittance can be combined with low thermal emittance and electrical conductivity irrespectively of the temperature [5]. The latter study also considers recent work on VO_2:Mg and shows that this dopant can significantly lower the luminous absorption in VO_2 [6].

POLARONIC ELECTROCHROMISM: A POLYMER ELECTROLYTE WITH NEAR-INFRARED-ABSORBING ITO NANOPARTICLES

EC "smart" windows normally have a centrally positioned electrolyte, which can be an inorganic thin film or a polymer electrolyte layer; this electrolyte connects two EC thin films—ideally one coloring under ion insertion and another coloring under ion extraction—and this three-layer stack is surrounded by transparent electrical conductors [7-9]. Reversible and persistent transmittance modulation takes place when a voltage is applied to the transparent electrical conductors. The physical mechanism dominating the EC effect is generally believed to be polaronic and involve electron transfer between neighboring transition metal ions [7,8]. The polymer-based device allows long open-circuit memory and uniform color changes, and it is also advantageous with regard to large-scale manufacturing [9]. Earlier work of ours considered a PEI-LiTFSI model electrolyte and studied ion relaxation mechanisms [10,11], mechanical properties [12], and SiO_2 nanoparticles for enhancing the ionic conductivity without compromising the optical performance [13]. For practical uses of "smart" windows it is often suitable to minimize the NIR part of the solar transmittance T_{sol} (in the $700 < \lambda < 3000$ nm wavelength range) independently of the luminous transmittance T_{lum} (at $400 < \lambda < 700$ nm).

The present work employed PEI-LiTFSI with maximum ion conductivity [10,11]. ITO nanoparticles with a Sn/In ratio of 5 at% were prepared though collaboration with the Lawrence Berkeley National Laboratory [3,14]. Nanocomposite electrolytes of (PEI-ITO)-LiTFSI were made with up to 7 wt% of ITO. Transmission electron microscopy applied to colloidal ITO nanoparticles (prior to a ligand exchange that was part of the preparation procedure) showed diameters D of 13 ± 3 nm. The spectral absorbance for such colloidal ITO nanoparticles displayed a strong peak at $\lambda \approx 1830$ nm.

Temperature-dependent ionic conductivity $\sigma(\tau)$ was recorded by impedance spectroscopy in the frequency interval $10^{-1} - 10^{7}$ Hz and at $23 < \tau < 70$ °C, as discussed before [10-13]. Figure 1 reports σ at six temperatures for samples with 0, 1, 3, 5 and 7 wt% ITO and shows that the ionic conductivity is almost independent of the ITO content and increases monotonically from $\sim 1\times10^{-6}$ S/cm at 23 °C to $\sim 2\times10^{-5}$ at 70 °C.

Spectral total and diffuse transmittance and reflectance were recorded at $300 < \lambda < 2500$ nm for 70 ± 10 μm thick (PEI-ITO)-LiTFSI layers. The diffuse spectral components were < 1 % in transmittance and < 0.3 % in reflectance and hence negligible. The solid curves in Fig. 2 show spectral transmittance $T(\lambda)$. Without ITO, the transmittance is large for $\lambda < 1500$ nm, and absorption bands due to PEI can be seen at $\lambda > 1500$ nm. Adding ITO produces a strong drop of the NIR transmittance while T_{lum} remains high. This is due to an absorption band whose intensity depends on the amount of ITO. Quantitatively, T_{lum} drops marginally from 91.7 to 83.3 % as the ITO content goes from zero to 7 wt% while the corresponding decrease of T_{sol} is as large as 88.9 to 56.3 %. These values were obtained from

$$T_{lum,sol} = \int d\lambda\, \varphi_{lum,sol}(\lambda)\, T(\lambda) \,/\, \int d\lambda\, \varphi_{lum,sol}\,, \qquad (1)$$

where φ_{lum} is the sensitivity of the light-adapted human eye [15] and φ_{sol} is solar irradiance for air mass 1.5 (the sun at 37° above the horizon) [16].

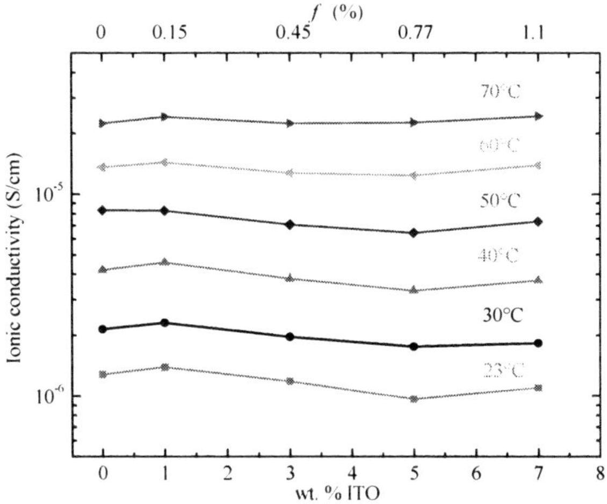

Figure 1. Ion conductivity vs ITO content in wt% and volume fraction f at several temperatures. Symbols indicate experimental data and connecting lines were drawn for convenience.

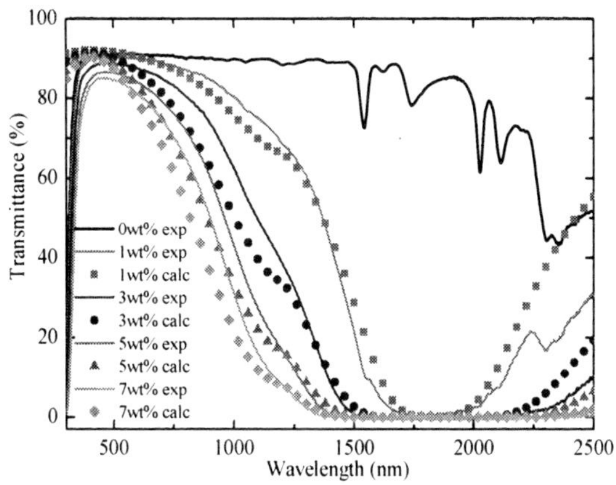

Figure 2. Spectral transmittance for the shown ITO contents as measured (solid curves) and calculated (symbols) for ~70-μm-thick (PEI-ITO)-LiTFSI layers.

The optical data in Fig. 2 could be interpreted by use of effective medium theory since $D \ll \lambda$. The formulations by Maxwell-Garnett [17] and Bruggeman [18] are frequently employed, and predictions from them coincide in the dilute limit which is applicable here. The latter theory yields an effective dielectric function, denoted ε^{Br}, from

$$f\,\frac{\varepsilon_{ITO}-\varepsilon^{Br}}{\varepsilon_{ITO}+2\varepsilon^{Br}}+(1-f)\frac{\varepsilon_m-\varepsilon^{Br}}{\varepsilon_m+2\varepsilon^{Br}}=0\ , \qquad (2)$$

where ε_{ITO} and $\varepsilon_m = 2.25$ are the dielectric functions of ITO and the polymer electrolyte, respectively, and $f \equiv f_{ITO}$ is the "filling factor" (*i.e.*, volume fraction) of ITO. The calculation of ε_{ITO} was carried out as discussed in detail in prior work [19] and presumed a free electron plasma with singly charged tin impurities screened according to the random phase approximation [20] with exchange represented by Hubbard's approach [21].

We first modeled the spectral absorbance data, mentioned above, using an empirical free electron density n_e—which is an essential fitting parameter in the theory for ε_{ITO}—of 0.98×10^{21} cm^{-3}. This value, which is of the expected magnitude [1], was then used to represent the data in Fig. 2. It is evident that theory and experiment are in good semi-quantitative agreement for all samples.

PLASMONIC ELECTROCHROMISM: APPROXIMATE PERFORMANCE LIMITS

A novel approach to electrochromism, which is different from the one discussed above, was presented recently by Milliron *et al.* [22,23], who demonstrated that transmittance modulation—especially in the NIR—can be obtained electrochemically in layers with nanoparticles consisting of a transparent and electrically conducting material such as ITO. This opens possibilities to use plasmonic effects in order to achieve a significant modulation of T_{sol} while T_{lum} remains high.

The conduction electrons in ITO form a free electron plasma with a density n_e and a (screened) plasma frequency ω_p at

$$\omega_p \approx [n_e e^2/(\varepsilon_0 \varepsilon_\infty m_c^*)]^{1/2}\ , \qquad (3)$$

where e is the electron charge, ε_0 is the permittivity of free space, ε_∞ is the high-energy dielectric constant, and m_c^* is the electrons' effective conduction-band mass. ITO films can have $n_e \approx 2.5 \times 10^{21}$ cm^{-3} [24] and similar n_es have been inferred for ZnO:Al and ZnO:Ga [24]. This makes the plasmon resonance at ω_p lie in the NIR or red end of the luminous spectrum. Specifically, the interesting recent discovery by Milliron *et al.* [22,23] is that the plasmon frequency of ITO and ZnO:Al nanoparticles can be persistently and reversibly tuned by electrochemical post-treatment, which modulates n_e most probably via the creation of dynamic accumulation and depletion layers at the nanoparticles' surfaces. Thus the plasma absorption can be swept through the NIR so that a variable part of the solar radiation is absorbed while the luminous absorption is influenced to a much smaller degree.

Calculations of the optical properties of ITO nanoparticles in a dielectric medium can be performed in the same way as above, and the upper panel of Fig. 3 illustrates spectral absorptance for a 5-μm-thick layer of ITO nanoparticles in a dielectric medium with $\varepsilon_m = 2.25$ (representative of typical solvents, polymers and glasses). Clearly the peak shifts from long wavelengths to ~1.3 μm when n_e is increased to 2×10^{21} cm^{-3}. The lower panel of Fig. 3 shows absorptance for $n_e = 1 \times 10^{21}$ cm^{-3} and layer thicknesses of $1 \leq d \leq 100$ μm and indicates that an interval of high absorptance covers a large part of the NIR for the largest values of d while the luminous absorptance remains low.

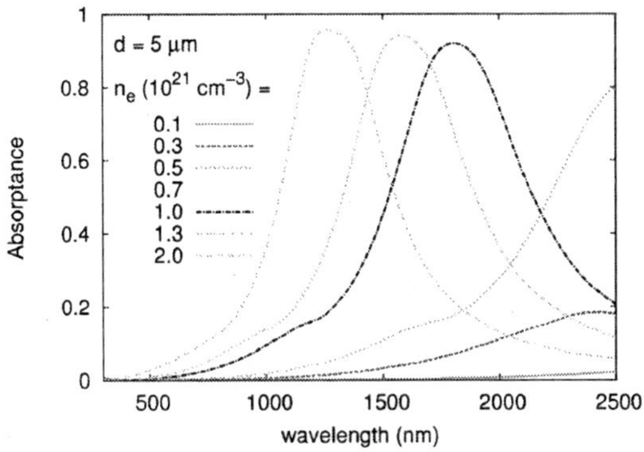

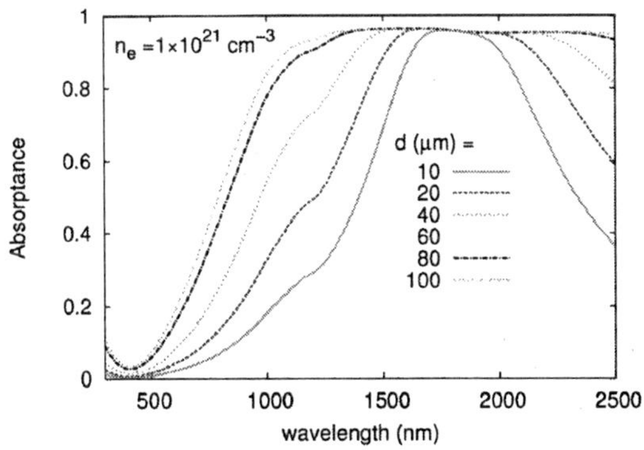

Figure 3. Calculated spectral absorptance for a 5-μm-thick layer with electron densities of $0.1 \times 10^{21} \leq n_e \leq 2 \times 10^{21}$ cm^{-3} (upper panel) and for layers with thicknesses of $10 \leq d \leq 100$ μm and $n_e = 1 \times 10^{21}$ cm^{-3} (lower panel). The layer contains 1 vol% of dispersed spherical ITO nanoparticles.

Quantitative data on $T_{\mathrm{lum}}(d,n_e)$ and $T_{\mathrm{sol}}(d,n_e)$ were obtained from Eq. (1) and are summarized in Fig. 4, showing that for each T_{lum} there is a well-defined minimum solar transmittance, denoted $T_{\mathrm{sol,min}}$. The relationship can be fitted to an empirical expression, given by the curve in Fig. 4, by $T_{\mathrm{sol,min}} \approx a + bT_{\mathrm{lum}} + c(T_{\mathrm{lum}})^2 + d(T_{\mathrm{lum}})^3$ with $a = -1.03$, $b = 5.94$, $c = -9.26$, and $d = 5.27$, which gives an approximate performance limit for a smart window with plasmon-induced electrochromism based on transparent conducting nanoparticles.

The generality and accuracy of the performance limits are considered next. The same theories as for ITO can be applied to transparent conductors based on ZnO [25] and SnO_2 [26], implying there is a certain freedom regarding choice of material. The limitation to a dilute composite of nanoparticles is important, however, since otherwise particle-particle interaction widens the plasmon absorption compared to the data in Fig. 3 and strong absorption in the NIR tends to extend also into the luminous range. Similar broadening occurs if nanoparticle clustering or non-spherical shapes are prevalent. The maximum layer thickness was set to 100 μm, which is arbitrary, and by going to much thicker layers the values of $T_{sol,min}$ can be 0.01 to 0.02 lower than those in Fig. 4 for a given T_{lum}. Furthermore optical properties of the layers are taken to be isotropic; however highly anisotropic properties—for example aligned nanorods—might yield slightly improved performance limits, and the same would occur for a judicious choice of nanoparticles with a range of electron densities.

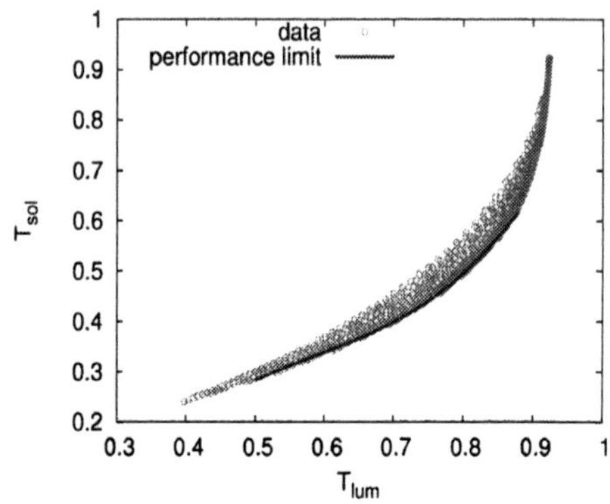

Figure 4. Calculated luminous and solar transmittance (T_{lum} and T_{sol}, respectively). The curve delineates an approximate performance limit for plasmonic EC devices.

THERMOCHROMICS: ITO-VO_2 NANOCOMPOSITES WITH LOW THERMAL EMITTANCE

Vanadium dioxide is a TC material of large interest for energy efficient fenestration and has been discussed in this context for many years [1,24,27]. Bulk VO_2 is monoclinic and semiconducting for $\tau < \tau_c$, where $\tau_c \approx 68$ °C, and is tetragonal and metallic-like at $\tau > \tau_c$ [28]. For practical applications it is obvious that (*i*) τ_c should be decreased to a value close to room temperature, (*ii*) the film must have a high enough T_{lum}, and (*iii*) the decrease of T_{sol} should be large enough when τ_c is exceeded. The latter parameter, denoted ΔT_{sol}, is defined by

$$\Delta T_{sol} \equiv T_{sol}(\tau < \tau_c) - T_{sol}(\tau > \tau_c) \quad . \tag{4}$$

The first of the requirements can be met by doping with tungsten and some other metals [24,27,29], but the latter two requirements have remained problematic for decades. However recent work of ours has shown by computation that a VO_2-based material consisting of nanoparticles, rather than being a thin film, increases T_{lum} as well as ΔT_{sol} and may make thermochromics promising for windows in buildings [27,30,31].

Coatings combining low thermal emittance E_{therm} with thermochromism independently of the temperature have rarely been discussed in detail, although recent work has reported data on VO_2 films with a top layer of an extremely thin layer of Pt [32] or ZnO:Al [33]. The best films of this type produced a modest combination of $T_{\text{lum}} \approx 46$ %, $\Delta T_{\text{sol}} \approx 4$ %, and $E_{\text{therm}} \approx 0.3$. Other approaches are possible [5], though, and here we report on a heavily doped wide band gap semiconductor incorporating VO_2 nanoparticles. We use ITO, as in the work on electrochromics discussed above, which can yield an excellent combination of high T_{lum} and low E_{therm} [19].

We consider a random mixture of nanoparticles of VO_2 and ITO and use again Bruggeman theory [18] to get the effective dielectric function ε^{Br} from

$$f\,\frac{\varepsilon_{VO2}-\varepsilon^{Br}}{\varepsilon_{VO2}+2\varepsilon^{Br}}+(1-f)\frac{\varepsilon_{ITO}-\varepsilon^{Br}}{\varepsilon_{ITO}+2\varepsilon^{Br}}=0\quad, \tag{5}$$

with $f \equiv f_{VO2}$. Data on ε_{VO2} were taken from previous work on sputter deposited 50-nm-thick VO_2 films [34]. Results for ε_{ITO} were based on measurements on high-quality ITO films with $n_e = 6.2 \times 10^{20}$ cm^{-3} [19]. These data are practically indistinguishable from the theoretical ones for ε_{ITO} used above.

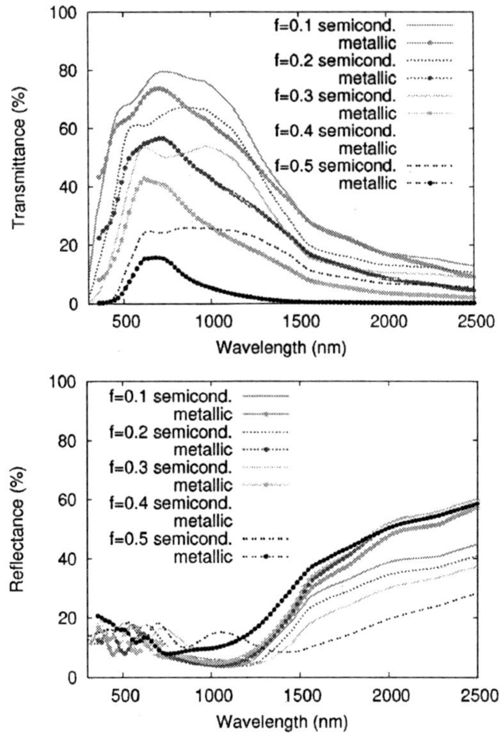

Figure 5. Calculated spectral transmittance and reflectance for ITO-VO_2 composites. The VO_2 volume fraction is denoted f. Data are given for $\tau < \tau_c$ (semiconducting VO_2) and $\tau > \tau_c$ (metallic VO_2).

Spectral and temperature-dependent transmittance and reflectance were computed from ε^{Br} at $300 < \lambda < 2500$ nm, and Fig. 5 shows data for composites of VO_2 and ITO wherein the ITO mass thickness was kept at 200 nm and the VO_2 mass thickness was varied so that the composite had filling factors in the $0.1 < f < 0.5$ range. The ITO nanoparticles are then percolating and their combined effect resembles that of an ITO film, which has a moderately strong plasma absorption in the NIR [19]. Thermochromism is manifest in the entire spectral interval, and the reflectance increases monotonically for $\lambda > 1000$ nm and reaches ~60 % at $\lambda = 2500$ nm for $\tau > \tau_c$. Data for $\tau < \tau_c$ and for films with low filling factors are consistent with earlier evaluations of spectral reflectance for 200-nm-thick ITO films [19]; the latter films showed $E_{therm} \approx 0.2$. Adding VO_2 enhances E_{therm} but the composite films still behave as low-emittance coatings, particularly for small VO_2 contents.

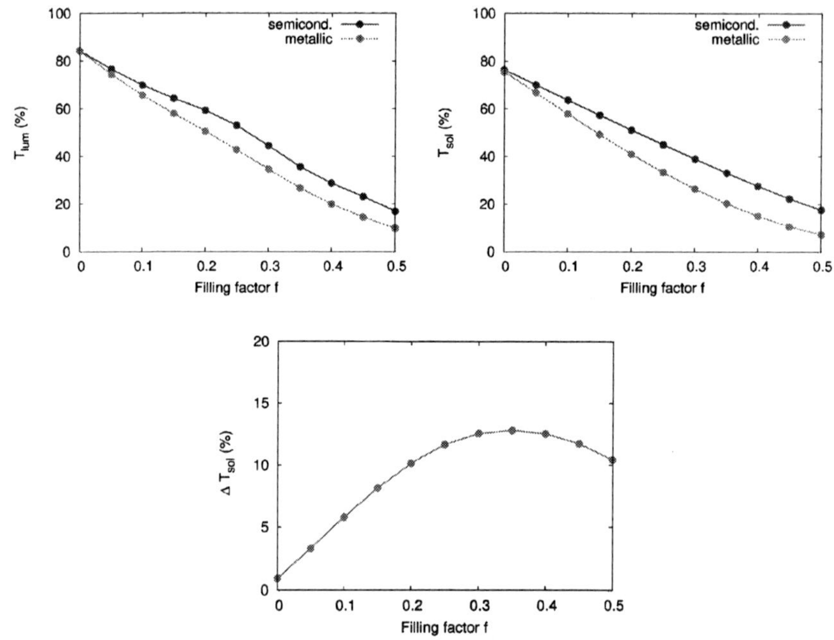

Figure 6. Calculated luminous and solar transmittance (T_{lum} and T_{sol}, respectively) and TC modulation of the solar energy transmittance (ΔT_{sol}) vs VO_2 fraction (f) for ITO-VO_2 composites. Data are shown for $\tau < \tau_c$ (semiconducting VO_2) and $\tau > \tau_c$ (metallic VO_2). Circles indicate data and connecting lines were drawn for convenience.

Luminous and solar transmittance were derived from Eq. (1), and the modulation of solar energy transmittance was obtained from Eq. (3). Figure 6 reports T_{lum} and T_{sol} for $\tau < \tau_c$ and $\tau > \tau_c$ as well as ΔT_{sol} vs VO_2 content. Expectedly, T_{lum} and T_{sol} decrease monotonically when f is increased. The integrated transmittance is higher for $\tau < \tau_c$ than for $\tau > \tau_c$ and the differences are similar for T_{lum} and for T_{sol}. The latter observation implies that thermochromism is not confined to the NIR but is significant in the full solar range. It is seen that $\Delta T_{sol} = 10$ % occurs at $f \approx 0.196$ and is consistent with $T_{lum} \approx 59.7$ % at $\tau < \tau_c$ and $T_{lum} \approx 51.1$ % at $\tau > \tau_c$. A broad maximum of ΔT_{sol} lies at ~12.8 % for $f \approx 0.35$.

Figure 5 shows that the transmittance drops sharply at $\lambda < 500$ nm. This feature of VO_2 ensues from the main optical band gap being undesirably low. Mg doping can improve the situation, however, as demonstrated a few years ago [34]. Figure 7 illustrates recent data [6] on $(\alpha\hbar\omega)^{1/2}$ vs photon energy $\hbar\omega$, where α is absorption coefficient, for films with the shown Mg contents and assuming indirect-allowed transitions. Extrapolations indicate that pure VO_2 has band gaps E_{g1} and E_{g2} at 1.67 eV and ~0.5 eV, respectively, which are consistent with well established literature data [35-37]. It is important to note that E_{g1} shifts towards higher energy for increasing doping, specifically from 1.67 eV to 2.32 eV as the atomic ratio Mg/(V + Mg) goes from zero to 0.190. The strong displacement of E_{g1} leads to a significantly decreased luminous absorptance, as inferred from the bell-shaped shaded region in Fig. 7 illustrating the relative luminous efficiency of the eye [15]. A 50-nm-thick VO_2-based film showed an increase of T_{lum} of ~10 % when the Mg content was ~7 at% [34].

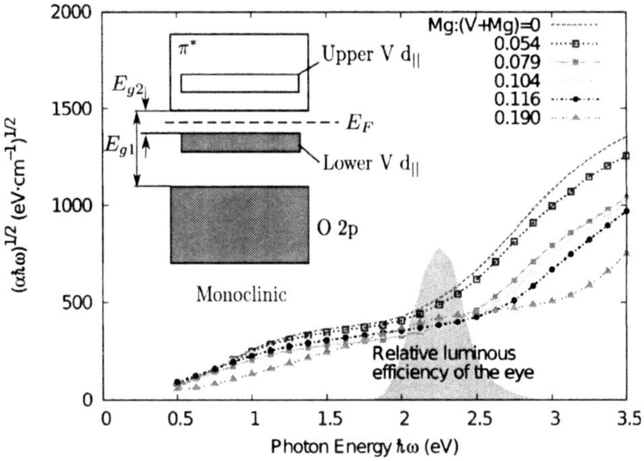

Figure 7. Main part shows $(\alpha\hbar\omega)^{1/2}$ vs photon energy $\hbar\omega$, where α is absorption coefficient, for Mg-doped VO_2 with the shown Mg/(V + Mg) atomic ratios. Inset illustrates the schematic band structure for monoclinic, semiconducting VO_2 according to Goodenough [29]. E_F denotes Fermi level. The shaded bell-shaped region signifies the relative luminous efficiency of the eye [15].

The observation of two band gaps can be understood from Goodenough's work on the electronic structure of VO_2 [29] as illustrated in the inset of Fig. 7. A crystal-field-induced splitting of the V3d t_{2g} band into $d_{\parallel}$ and π^* sub-bands, and dimerization of the V atoms in the otherwise tetragonal lattice, separates the $d_{\parallel}$ band into two parts. The lower $d_{\parallel}$ band is filled and the π^* band is empty, and E_{g1} is the band gap between O2p and π^* and E_{g2} the band gap between the lower $d_{\parallel}$ and π^*. This assignment of the band gaps cannot explain the effect of Mg, but hybrid functional calculation could reproduce the principal optical effect being an increased T_{lum} as the Mg fraction was increased [6]. Mg doping led to a decrease of τ_c [34] though not to the same degree as tungsten doping would produce.

SUMMARY AND CONCLUSIONS

This paper summarized several recent studies of ours [3-6] regarding new uses of ITO nanoparticles in electrochromics and thermochromics. Concerning the first topic, we incorporated ITO nanoparticles in the transparent polymer electrolyte of a polaronic EC device and showed that high haze-free luminous transmittance could be accomplished along with strong near-infrared absorption without deteriorating the ionic conductivity. Furthermore we showed that reliable performance limits could be stated for plasmonic EC devices based on ITO. Concerning the second topic, thermochromism, we showed that good luminous transmittance, low thermal emittance, and substantial thermochromic modulation of the solar energy throughput could be achieved in films comprising a mixture of ITO and a VO_2-based material. It was also shown that Mg-doping of VO_2 led to a decrease of the luminous absorptance which otherwise tends to be an obstacle for thermochromic fenestration.

It should be emphasized that our work is exploratory, and there are many development options and alternative transparent conducting nanoparticles such as ZnO:Al [23], SnO_2:Sb [38], M_xWO_3 (with M being Na, K, or Cs) [39-40], and LaB_6 [41-44]. There is also a huge number of techniques for nanoparticle and nanowire fabrication, as recently surveyed with regard to ITO [45].

ACKNOWLEDGEMENTS

The ITO nanoparticles included in the polymer electrolyte were prepared through collaboration with E. L. Runnerstrom and D. J. Milliron at the Molecular Foundry of the Lawrence Berkeley National Laboratory, Berkeley, CA, USA. Financial support was received from the Swedish Research Council (VR) and the Swedish Research Council for Environment, Agricultural Sciences, and Spatial Planning (FORMAS).

REFERENCES

1. G. B. Smith and C. G. Granqvist, *Green Nanotechnology: Solutions for Sustainability and Energy in the Built Environment* (CRC Press, Boca Raton, 2010).
2. C. M. Lampert and C. G. Granqvist, *Large-Area Chromogenics: Materials and Devices for Transmittance Control* (SPIE – The International Society for Optical Engineering, Bellingham, WA, 1990), SPIE Institutes for Advanced Optical Technologies, Vol. IS4.
3. İ. Bayrak Pehlivan, E. L. Runnerstrom, S.-Y. Li, G. A. Niklasson, D. J. Milliron and C. G. Granqvist, *Appl. Phys. Lett.* **100**, 241902 (2012).

4. S.-Y. Li, G. A. Niklasson and C. G. Granqvist, *Appl. Phys. Lett.* **101**, 071903 (2012).
5. S.-Y. Li, G. A. Niklasson and C. G. Granqvist, *Appl. Phys. Lett.* **99**, 131907 (2011).
6. S. Hu, S.-Y. Li, R. Ahuya, C. G. Granqvist, K. Hermansson, G. A. Niklasson and R. H. Scheicher, Appl. Phys. Lett. **101**, 201902 (2012).
7. C. G. Granqvist, *Handbook of Inorganic Electrochromic Materials* (Elsevier, Amsterdam, The Netherlands, 1995).
8. G. A. Niklasson and C. G. Granqvist, *J. Mater. Chem.* **17**, 127 (2007).
9. C. G. Granqvist, *Sol. Energy Mater. Sol. Cells* **99**, 1 (2012).
10. İ. Bayrak Pehlivan, R. Marsal, P. Georén, C. G. Granqvist and G. A. Niklasson, *J. Appl. Phys.* **108**, 074102 (2010).
11. İ. Bayrak Pehlivan, P. Georén, R. Marsal, C. G. Granqvist and G. A. Niklasson, *Electrochim. Acta* **57**, 201 (2011).
12. İ. Bayrak Pehlivan, R. Marsal, C. G. Granqvist, G. A. Niklasson and P. Georén, *Sol. Energy Mater. Sol. Cells* **94**, 2399 (2010).
13. İ. Bayrak Pehlivan, C. G. Granqvist, R. Marsal, P. Georén and G. A. Niklasson, *Sol. Energy Mater. Sol. Cells* **98**, 465 (2012).
14. A. Llordes, A. T. Hammack, R. Buonsanti, R. Tangirala, S. Aloni, B. A. Helms and D. J. Milliron, *J. Mater. Chem.* **21**, 11631 (2011).
15. G. Wyszecki and W. S. Stiles, *Color Science: Concepts and Methods, Quantitative Data and Formulae*, 2nd ed. (Wiley, New York, 2000).
16. ASTM G173-03 Standard Tables of Reference Solar Spectral Irradiances: Direct Normal and Hemispherical on a 37° Tilted Surface, *Annual Book of ASTM Standards* (American Society for Testing and Materials, Philadelphia, PA, 2008); Vol. 14.04.
17. J. C. Maxwell-Garnett, *Phil. Trans. R. Soc. London, Ser. A*, **203**, 385 (1904); **205**, 237 (1906).
18. D. A. G. Bruggeman, *Ann. Phys. (Leipzig)* **24**, 636 (1935).
19. I. Hamberg and C. G. Granqvist, *J. Appl. Phys.* **60**, R123 (1986).
20. J. Lindhard, *Kgl. Danske Videnskab. Selskab Mat.-Fys. Medd.* **28**, No. 8 (1954).
21. J. Hubbard, *Proc. R. Soc. London, Ser. A*, **243**, 336 (1957).
22. G. Garcia, R. Buonsanti, E. L. Runnerstrom, R. J. Mendelsberg, A. Llordes, A. Anders, T. J. Richardson and D. J. Milliron, *Nano Lett.* **11**, 4415 (2011).
23. R. Buonsanti, A. Llordes, S. Aloni, B. A. Helms and D. J. Milliron, *Nano Lett.* **11**, 4706 (2011).
24. C. G. Granqvist, *Sol. Energy Mater. Sol. Cells* **91**, 1529 (2007).
25. Z.-C. Jin, I. Hamberg and C. G. Granqvist, *J. Appl. Phys.* **64**, 5117 (1988).
26. B. Stjerna, E. Olsson and C. G. Granqvist, *J. Appl. Phys.* **76**, 3797 (1994).
27. S.-Y. Li, G. A. Niklasson and C. G. Granqvist, *Thin Solid Films* **520**, 3823 (2012).
28. F. J. Morin, *Phys. Rev. Lett.* **3**, 34 (1959).
29. J. B. Goodenough, *J. Solid State Chem.* **3**, 490 (1971).
30. S.-Y. Li, G. A. Niklasson and C. G. Granqvist, *J. Appl. Phys.* **108**, 063525 (2010).
31. S.-Y. Li, G. A. Niklasson and C. G. Granqvist, *J. Appl. Phys.* **109**, 113515 (2011).
32. L. Kang, Y. Gao, Z. Chen, J. Du, Z. Zhang and H. Luo, *Sol. Energy Mater. Sol. Cells* **94**, 2078 (2010).
33. L. Kang, Y. Gao, H. Luo, J. Wang, B. Zhu, Z. Zhang, J. Du, M. Kanehira and Y. Zhang, *Sol. Energy Mater. Sol. Cells* **95**, 3189 (2011).
34. N. R. Mlyuka, G. A. Niklasson and C. G. Granqvist, *Appl. Phys. Lett.* **95**, 171909 (2009).

35. H. W. Verleur, A. S. Barker, Jr., and C. N. Berglund, *Phys. Rev.* **172**, 788 (1968).
36. M. M. Qazilbash, K. S. Burch, D. Whisler, D. Shrekenhamer, B. G. Chae, H. T. Kim and D. N. Basov, *Phys. Rev. B* **74**, 205118 (2006).
37. Z. Luo, Z. Wu, X. Xu, T. Wang and Y. Jiang, *J. Vac. Sci. Technol. A* **28**, 595 (2010).
38. T. G. Conti, A. J. Chiquito, R. A. da Silva, E. Longo and E. R. Leite, *J. Am. Ceram. Soc.* **93**, 3862 (2010).
39. A. Adachi and T. Asahi, *J. Mater. Res.* **27**, 965 (2012).
40. C. Guo, S. Yin and T. Sato, *J. Am. Ceram. Soc.* **95**, 1634 (2012).
41. S. Schelm and G. B. Smith, *Appl. Phys. Lett.* **82**, 4346 (2003).
42. S. Schelm, G. B. Smith, P. D. Garrett and W. K. Fisher, *J. Appl. Phys.* **97**, 124314 (2005).
43. H. Takeda, H. Kuno and K. Adachi, *J. Am. Ceram. Soc.* **91**, 2897 (2008).
44. K. Adachi, M. Miratsu and T. Asahi, *J. Mater. Res.* **25**, 510 (2010).
45. E. N. Dattoli and W. Lu, *MRS Bull.* **36**, 782 (2011).

Mater. Res. Soc. Symp. Proc. Vol. 1492 © 2013 Materials Research Society
DOI: 10.1557/opl.2013.591

Photochromic properties in Eu^{3+} doped Sr_2SnO_4

Sunao Kamimura[1] Hiroshi Yamada[1,2] and Chao-Nan Xu [1,2,3]
[1] Interdisciplinary Graduate School of Engineering Science, Kyushu University, Fukuoka 816-8580, Japan
[2] National Institute of Advanced Industrial Science and Technology (AIST), Saga 841-0052, Japan
[3] International Institute for Carbon-Neutral Energy Research (WPI-I2CNER), Fukuoka 819-0395, Japan

ABSTRACT

We report the photochromism (PC) in Eu^{3+} doped Sr_2SnO_4. The Sr_2SnO_4:Eu^{3+} is sensitive to UV light ($\lambda < 350$ nm) and turns to purple color. When visible light (λ= 400-700 nm) is irradiated, the colored Sr_2SnO_4:Eu^{3+} is bleached. From the results of absorption spectrum, Sr vacancies in the host lattice and charge transfer transition between Eu^{3+}-O^{2-} is responsible for PC process of Sr_2SnO_4:Eu^{3+}.

INTRODUCTION

In general, perovskite oxides show various functional properties such as switchable polarization, piezoelectricity, pyroelectricity, and non-linear dielectric behavior [1]. These properties are indispensable for application in electric and optical devices such as non-volatile memories, thermoelectric modules, sensors, microactuators, infrared detectors, microwave phase filters, and so on [1-3]. Despite the diverse function of perovskite oxides, photochromic (PC) properties had been little known until in 1970s Faughnan reported transition-metal doped $SrTiO_3$ [4], which turns to black color by visible light irradiation and the black color is bleached by heating. Here PC is a reversible change in color of materials under light irradiation, and it has application that include rewritable copy papers, smart windows, optical filters and photoswitchs [5,6]. Especially, inorganic PC materials are expected as the high density optical memory devices with small semiconductor laser diodes due to their excellent thermal stability and chemical resistance [7]. Although some inorganic oxides, such as WO_3, MoO_3, TiO_2 and $BaMgSiO_4$, display PC performance [8-11], perovskite oxides which exhibit PC are very attractive because they can be used as multi-functional materials.

Recently, rare-earth doped Sr_2SnO_4, which belongs to the layered perovskite-related structure with space group *Pbnb*, lattice parameters a = 5.735, b = 12.587, c = 5.728 Å and V = 413.55 Å^3 [12], has attracted attention due to showing various optical properties, such as photoluminescence, mechanoluminescence, upconversion [13,14].

In this paper, we report PC performance in Eu^{3+} doped Sr_2SnO_4. The Sr_2SnO_4:Eu^{3+} shows a colorless before light irradiation and then, after UV light ($\lambda < 350$ nm) irradiation, it turns to be purple color. After visible light (λ= 400-700 nm) irradiation, the colored Sr_2SnO_4:Eu^{3+} is bleached. The PC process for Sr_2SnO_4:Eu^{3+} is discussed in accordance with their absorption properties and crystal structure.

EXPERIMENTAL

The Sr_2SnO_4:Eu^{3+} powders were synthesized by a solid reaction process as follows. The starting materials of ultrafine powder of $SrCO_3$, Eu_2O_3, SnO_2 were weighted according to the chemical stoichiometry of $Sr_{1.995}Eu_{0.005}SnO_{4-\delta}$ and mixed together thoroughly with a small amount of ethanol. After drying, the mixture was pre-fired at 1073 K for 1 h in air and then mixed thoroughly again. The mixed powder was pressed into disks and fired at 1873 K for 5 h in air. The crystalline phase of $Sr_2SnO_4:Eu^{3+}$ was characterized with a powder X-ray diffraction instrument (RINT-2000, Rigaku Co., Japan) with CuKα (1.5418 Å) radiation (cathode voltage 40 kV and current 40 mA). The XRD pattern for $Sr_2SnO_4:Eu^{3+}$ was coincident with the PDF standard of the orthorhombic phase of Sr_2SnO_4 (JCPDS 00-052-1590), and no other impurity phases could not be detected, demonstrating that single phase of the compound was synthesized. The absorption spectrum was recorded with a UV/VIS/NIR spectrometer (V570, JASCO Co., Japan) at room temperature.

RESULTS AND DISCUSSION

Figure 1(a) shows the absorption spectra and pictures for $Sr_2SnO_4:Eu^{3+}$ colored and bleached. Firstly the $Sr_2SnO_4:Eu^{3+}$ did not absorb any visible light, corresponding colorless sample. When the UV light was irradiated on the pellet sample, the strong absorption peak centered at 578 nm was appeared and the sample color turned to purple color. It was conformed that the coloration was occurred at the irradiation wavelength $\lambda < 350$ nm. It is interesting that this absorption peak was disappeared and its purple color was immediately bleached after the visible light irradiation (λ= 400-700 nm). As shown in Fig. 1(b), the absorption intensity peaking at 578 nm was repeatedly recovered by the UV irradiation at least more than 10 times.

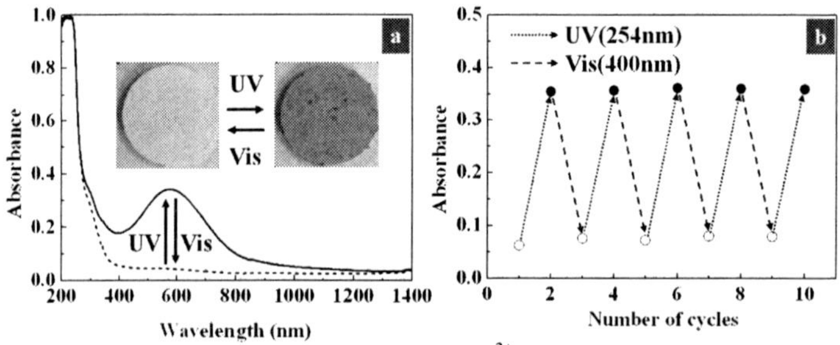

Figure 1. (a) The absorption spectra for $Sr_2SnO_4:Eu^{3+}$ and insets shows typical pictures of the photochromic performance. (b) Absorption changes of $Sr_2SnO_4:Eu^{3+}$ at 578 nm induced by alternating 254 nm (614μW/cm^2) and 400 nm (5620μW/cm^2) light irradiation.

Figure 2 shows the response of $Sr_2SnO_4:Eu^{3+}$ to UV (254 nm) and VIS (400 nm) light irradiation. In the response to UV light irradiation, the absorbance at 578nm was rapidly increased with time and it was reached the maximum at 60 s. In general, many inorganic PC materials show slow response: for example, $Sr_2TiO_4:Cr^{3+}$ and Ag-doped TiO_2 colored by light irradiation at one week and 15 minutes, respectively [8,15]. Therefore, this result indicated that $Sr_2SnO_4:Eu^{3+}$ is good candidate for application such as optical memory, rewritable copy paper. It should be noted that the purple colored $Sr_2SnO_4:Eu^{3+}$, if kept in the dark, has considerable coloration after 24 hours.

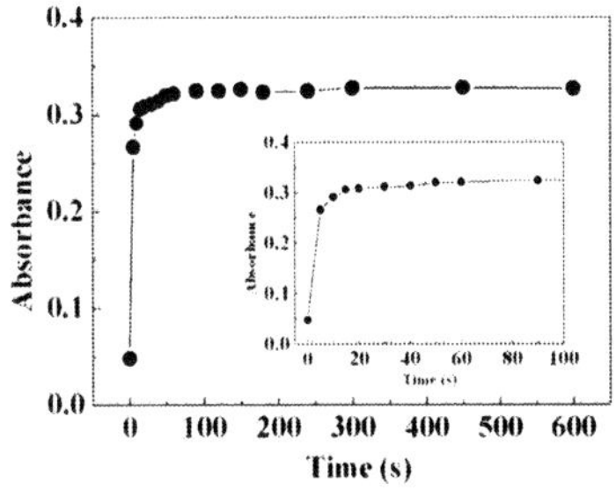

Figure 2. Response of $Sr_2SnO_4:Eu^{3+}$ to UV light irradiation (λ=254 nm : 614μW/cm^2). The inset shows a reproduction of a part of the same figure at a finer time resolution.

To understand the PC process for $Sr_2SnO_4:Eu^{3+}$, the absorption properties were further investigated. Figure 3(a) shows the comparison of the absorption spectra between non-doped Sr_2SnO_4 and Eu^{3+}-doped Sr_2SnO_4 before UV light irradiation in the UV region. Eu^{3+}-doped sample showed a small shoulder peak below the host absorption, while the non-doped sample didn't show that peak. This shoulder peak could be assigned to the charge transfer state (CTS) between O^{2-}-Eu^{3+} [16]. It is emphasized that non-doped sample didn't show the PC performance (refer Fig.2), implying that Eu^{3+}-CTS plays important roles for the PC properties. It is helpful to analyze the irradiation wavelength dependence of the absorption intensity at 578 nm to understand the relationship between CTS and PC phenomenon (see Fig. 3(b)). The absorbance for $Sr_2SnO_4:Eu^{3+}$ was strongly depended on the irradiation wavelength: the absorbance was gradually increased as shorter wavelength and it was found to have the maximum at 260 nm, which close to the band edge of host materials. However, taking into consideration that PC performance was clearly observed even where there is no absorption in host materials, the PC process is possibly considered to originate from the charge transfer absorption between O^{2-}-Eu^{3+} rather than the host absorption. Here it should be noted that greater than band-gap energy also produces a formation of Eu^{3+}-CTS which can lead to a PC phenomenon.

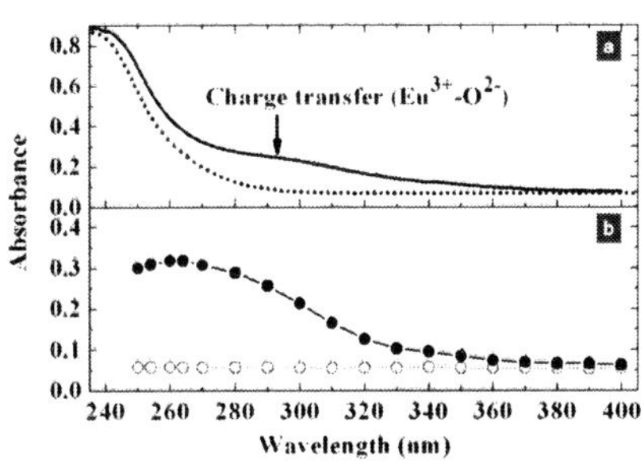

Figure 3. (a) Absorption spectra for Eu^{3+}-doped Sr_2SnO_4 (solid line) and non-doped Sr_2SnO_4 (dotted line). (b) Irradiation wavelength dependence of the absorption intensity at 578 nm for Eu^{3+}-doped Sr_2SnO_4 (black points) and non-doped Sr_2SnO_4 (white points).

The observed PC phenomenon can be understood on the basis of the carrier trapping model.[4] In this model the carriers of electrons and/or holes generated by photon absorption were captured into the lattice defects such as a point defect, an co-doped ion, and others, so-called color center. The color center absorbed the visible light, resulting in displaying to the various colors in materials. Thus lattice defects play an important role for the PC performance. In the case of $Sr_2SnO_4:Eu^{3+}$, supposing that Eu^{3+} ions are incorporated into Sr^{2+} sites [16], two point defects of the Eu impurity ($Eu^{\cdot}_{Sr}$) and the Sr vacancy (V''_{Sr}) were created in the host lattice due to charge compensation. Here $Eu^{\bullet}_{Sr}$ and V''_{Sr} act as the electron and hole trap sites, respectively.

On the basis of the above experimental results, the PC mechanism for $Sr_2SnO_4:Eu^{3+}$, which was divided to three processes, has been proposed as follows:

(a) The carrier trapping process; When UV light ($\lambda < 350$ nm) was irradiated, the electron transition was occurred from the O^{2-} to Eu^{3+}, forming to Eu^{3+}-CTS, where the electron was located around of Eu^{3+} ions like "Eu^{3+} plus electron". After that, the generated holes were partly transferred into the V''_{Sr} [17].

(b) The coloration process; That "Eu^{3+} plus electron" may act as a color center in $Sr_2SnO_4:Eu^{3+}$. The electron transition occurred in "Eu^{3+} plus electron" from the ground state to the excited state, corresponded to the absorption spectrum centered at 578 nm, resulting in displaying the purple color in material. It should be noted that luminescence from "Eu^{3+} plus electron" could not be observed.

(c) The bleaching process; When visible light (λ= 400-700 nm) was irradiated, trapped holes at V''_{Sr} were released. No luminescence from $Sr_2SnO_4:Eu^{3+}$ was observed upon visible light (λ= 400-700 nm) irradiation, implying that released holes were recombined non-radiatively with electron in "Eu^{3+} plus electron"[18], leads to displaying the colorless sample.

CONCLUSIONS

We have found the PC phenomenon in Eu^{3+} doped Sr_2SnO_4, which reversibly change color by alternate irradiation with UV ($\lambda < 350$ nm) and VIS (λ= 400-700 nm) light. Therefore, $Sr_2SnO_4:Eu^{3+}$ is a promising candidate for optical memories, ultraviolet photo detectors. Although the proposed model explains that the PC is intimately related to the lattice defects and the valence change of europium ions, other aspects such as the energy level of Eu^{3+} is not yet clear. Further research is required to improve the present model and understand the PC mechanism in depth.

ACKNOWLEDGMENTS

The authors thank their colleagues at National Institute of Advanced Industrial Science and Technology (AIST) for assistance and advise with regard to this study.

REFERENCES

[1] F. S. Galasso, *Structure, Properties and Preparation of Perovskite-type compounds* (Pergamon, Oxford, 1969).

[2] K. Koumoto, Y. Wang, R. Zhang, A. Kosuga, and R. Funahashi, *Annu. Rev. Mater. Res.* **40**, 363 (2010).

[3] X. S. Wang, C. N. Xu, H. Yamada, K. Nishikubo, and X. G. Zheng, *Adv. Mater.* **17**, 1254 (2005).

[4] B. W. Faughnan, and Z. J. Kiss, *Phys. Rev. Lett.* **21**. 1331 (1968).

[5] A. J. Cohen, and H. L. Smith, *Science*. **137**. 981 (1962).

[6] G. Poirier, M. Nalin, L. Cescato, Y. Messaddeq, and S. J. L. Rebeiro, *J. Chem. Phys.* **125**. 161101 (2006).
[7] J. N. Yao, K. Hashimoto, and A. Fujishima, *Nature*. **355**. 624 (1992).
[8] Y. Ohko, T. Tatsuma, T. Fujii, K. Naoi, C. Niwa, Y. Kubota, and A. Fujishima, *Nature. Mater*. **2**, 29 (2003).
[9] M. Akiyama, *Appl. Phys. Lett.* **97**, 181905 (2010).
[10] T. He, and J. Yao, *Prog. Mater. Sci.* **51**, 810 (2006).
[11] Z. G. Yi, H. Iwai, and J. H. Ye1, *Appl. Phys. Lett.* **96**. 114103 (2010).
[12] W. T. Fu, D. Visser, and D. J. W. IJdo, *J. Solid. State. Chem.* **169**, 208 (2002).
[13] S. Kamimura, H. Yamada, and C. N. Xu, *Appl. Phys. Lett.* **101**, 091113 (2012).
[14] Y. C. Chen, Y. H. Chang, and B. S. Tsai, *J. Alloys. Comp.* **398**, 256 (2005).
[15] Z. G. Yi, H. Iwai, and J. H. Ye1, *Appl. Phys. Lett.* **96**. 114103 (2010).
[16] H. M. Yang, J. X. Shi, and M. L. Gong, *J. Alloy. Compd.* **415**, 213 (2006).
[17] C. N. Xu, T. Watanabe, and M. Akiyama, and X. G. Zheng, *Appl. Phys. Lett.* **74**, 2414 (1999).
[18] C. W. Struck, and W. H. Fonger, *Phys. Rev. B.* **4**, 22 (1971).

Mater. Res. Soc. Symp. Proc. Vol. 1492 © 2013 Materials Research Society
DOI: 10.1557/opl.2013.592

Strong light emission from stress-activated perovskite-related oxides

Sunao Kamimura[1] Hiroshi Yamada[1,2] and Chao-Nan Xu[1,2,3 *]
[1] Interdisciplinary Graduate School of Engineering Science, Kyushu University, Fukuoka 816-8580, Japan
[2] National Institute of Advanced Industrial Science and Technology (AIST), Saga 841-0052, Japan
[3] International Institute for Carbon-Neutral Energy Research (WPI-I^{2}CNER), Kyushu University, Fukuoka 819-0395, Japan

ABSTRACT

We report mechanoluminescence (ML) in Sm^{3+}-doped $Sr_3Sn_2O_7$ phosphors with perovskite-related structure. The ML from $Sr_3Sn_2O_7$:Sm^{3+}, emitted strong reddish-orange light upon compression, is clearly observable by naked eyes. Based on the comparison between ML spectrum and photoluminescence spectrum, the ML emission was identified to be due to electron transition from an excited state $^4G_{5/2}$ to the ground state ^6H_J (J = 5/2, 7/2, 9/2) in Sm^{3+} ions. Although the ML emission was gradually decayed as compressive load was applied repeatedly, it recovered completely upon irradiation with UV light (254 nm). This behavior indicating that ML of $Sr_3Sn_2O_7$:Sm^{3+} is intimately related to the charge traps. The charge transfer state (CTS) band in the PL excitation spectra was observed for $Sr_3Sn_2O_7$:Sm^{3+}, indicating that the efficient energy transfer from the host to the Sm^{3+} ions. The formation of CTS and the charge traps may be responsible for this ML performance.

INTRODUCTION

Perovskite oxides show a wide range of functional properties such as switchable polarization, piezoelectricity, pyroelectricity, thermoelectricity, and nonlinear dielectric behavior. These properties are crucial for their use in electric devices, such as nonvolatile memory, sensors, microactuators, infrared detectors, and microwave phase filters [1,2]. Recently, Pr^{3+}-doped $Ba_{1-x}Ca_xTiO_3$ (x > 0.25) diphasic perovskite oxides have attracted attention as multi-functional materials, which show electroluminescence, electrostriction and mechanoluminescence [3]. Mechanoluminescence (ML) is one kind of phenomenon of luminescence where photons are emitted by mechanical stimuli; it is of particular interest for use in stress sensing devices which detect damage, fracture, and deformation in various structures [4-6]. Although some inorganic oxides, such as $SrAl_2O_4$:Eu^{2+} [7], $CaAl_2Si_2O_8$:Eu^{2+} [8], $SrMg_2(PO_4)_2$:Eu^{2+} [9], display ML, perovskite oxides which exhibit intense ML are very attractive because they can be used as multi-functional materials.

Recently, we have found ML in Sm^{3+}-doped $Sr_3Sn_2O_7$ phosphors [10], which consist of stacked perovskite double layers (n = 2; a = 5.710, b = 5.736, c = 20.688 Å, space group *Amam*) [11]. The ML from $Sr_3Sn_2O_7$:Sm^{3+}, emitted strong reddish-orange ML upon compression, is clearly observable by naked eyes. However, the ML mechanism of $Sr_3Sn_2O_7$:Sm^{3+} still not clear. In the present research, the ML properties of $Sr_3Sn_2O_7$:Sm^{3+} were further investigated and the ML process was discussed in conjunction with other optical properties, such as photoluminescence (PL), thermoluminescence and absorption properties.

*) Author to whom correspondence should be addressed. Electronic mail: cn-xu@aist.go.jp

EXPERIMENTAL

$Sr_3Sn_2O_7:Sm^{3+}$ powders were synthesized by a solid state reaction process. Ultrafine $SrCO_3$, Sm_2O_3, and SnO_2 powders in the appropriate proportions for $Sr_{2.99}Sm_{0.01}Sn_2O_{7-\delta}$, were mixed thoroughly with a small amount of ethanol. The mixtures were pre-fired at 1073 K for 1 h in air and then mixed thoroughly again. The mixed powder was pressed into a disk and fired at 1573 K for 5 h in air. The crystalline phase of $Sr_3Sn_2O_7:Sm^{3+}$ was characterized with a powder X-ray diffraction instrument (RINT-2000, Rigaku Co., Japan) with CuKα (1.5418 Å) radiation (cathode voltage 40 kV and current 40 mA). The ML properties were investigated at room temperature by using our previously reported lab-made measurement system [7-10]; compressive load was applied to the $Sr_3Sn_2O_7:Sm^{3+}$/epoxy composite pellet after it was exposed to UV light (254 nm) for 1 min. The PL and ML spectra were measured with a photon multichannel analyzer system (PMA-100, Hamamatsu Photonics Co., Japan). The PL excitation, emission spectra and thermoluminescence were measured using a fluorescence spectrometer (FP6600, JASCO Co., Japan) and the absorption spectrum was recorded with a UV/VIS/NIR spectrometer (V570, JASCO Co., Japan) at room temperature.

RESULTS AND DISCUSSION

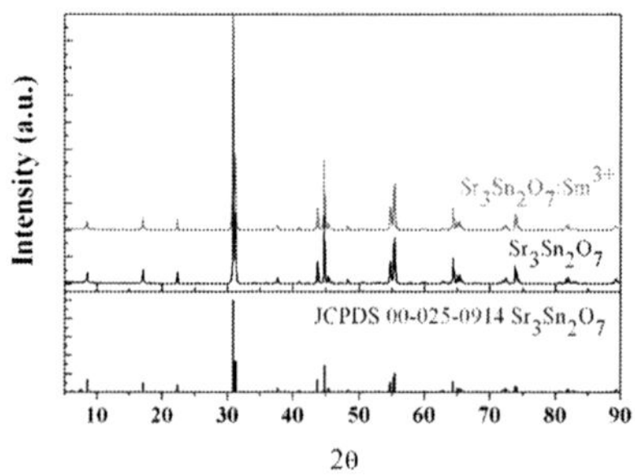

Figure 1. XRD patterns of synthesized $Sr_3Sn_2O_7:Sm^{3+}$ and $Sr_3Sn_2O_7$ powders and standard PDF data for $Sr_3Sn_2O_7$.

Figure 1 shows the X-ray diffraction patterns (XRD) for $Sr_3Sn_2O_7:Sm^{3+}$ and $Sr_3Sn_2O_7$. As seen in the figure, all the diffraction peaks coincide with those of the standard $Sr_3Sn_2O_7$ phase, with no other (impurity) phases being detectable.

Figure 2(a) shows a typical image of ML for $Sr_3Sn_2O_7:Sm^{3+}$ and Fig. 2(b) shows a typical ML response curves to two consecutive cycles of triangular loading sweep between 0 and 350 N,. A reddish-orange light could be observed clearly by naked eyes (>30 mcd/m^2), and the strongest light being emitted at the points of contact between the pellet and the testing machine. As seen in Fig. 2(b), in the first cycle, the ML increased almost linearly with increasing load, followed by a sharp decrease on releasing load. In the second cycle, similar response behavior was observed but with a much weaker intensity, indicating a decaying tendency of the ML with repeating loading tests. It should be noted that non-doped $Sr_3Sn_2O_7$ didn't show ML (see Fig. 2(b)), indicating that Sm^{3+} ion is important roles for ML performance.

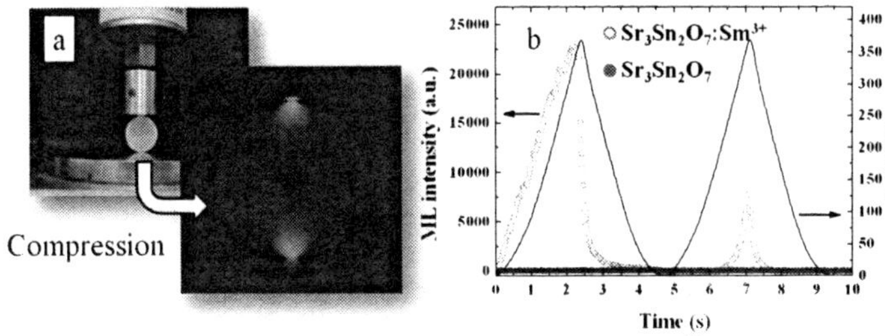

Figure 2. (a) The setting of the sample pellet ($Sr_3Sn_2O_7:Sm^{3+}$ composite) in a testing machine and an image of ML from the pellet put under a compressive load. (b) ML response of $Sr_3Sn_2O_7:Sm^{3+}$ (open circle) and $Sr_3Sn_2O_7$ (closed circle) composite pellet to triangular waves of loading.

It should be noted that the intensity could be completely recovered back once the pellet was irradiated with an ultraviolet light (254 nm) for 1min as shown in Figure 3, where the ML intensity of the pellet under the maximum compressive load applied (350 N) is shown as a function of the number of the loading test up to 5 cycles after the irradiation.

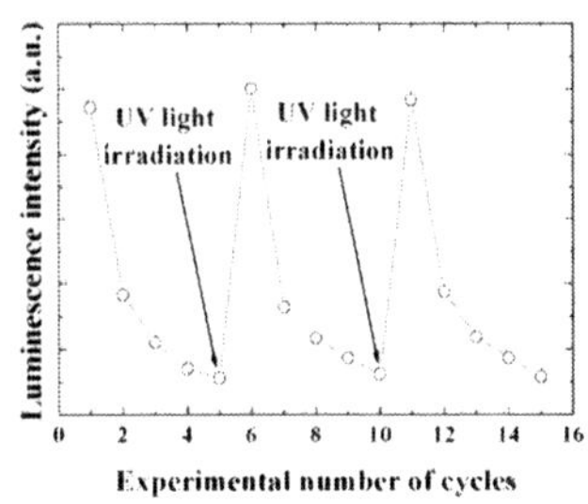

Figure 3. Decay of ML with repeating loading test and recovery with UV irradiation (254 nm).

The decay of luminescence with repeating tests followed by its complete recovery with an irradiation of the light, suggesting that ML phenomenon is intimately related to the charge traps [7-10]: an electron or hole trapped in a lattice defect, such as a point defect or co-doped ion, is released by a mechanical stimulus and recombines with a luminescence center, resulting in photon emission. Thus, charge traps play an important role in ML emission. In $Sr_3Sn_2O_7:Sm^{3+}$, assuming that the Sm^{3+} ions are incorporated into the Sr sites [12,13], two point defects arising from the Sm impurity ($Sm^{\cdot}_{Sr}$) and the Sr vacancy (V''_{Sr}) were created in the host lattice by charge compensation. Here $Sm^{\cdot}_{Sr}$ and V''_{Sr} act as the electron and hole trap sites, respectively.

In order to estimate their trap levels, the thermoluminescence (TL) measurement for $Sr_3Sn_2O_7:Sm^{3+}$ has been done from 273 K to 473 K at various heating rates (β) between 0.17 and 1.50 K / s. Resulting TL glow curves recorded at heating rates 1.5 K/s, are shown in Figure 4.

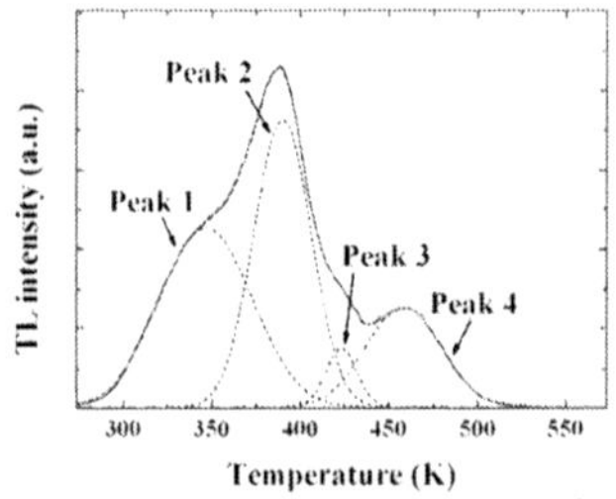

Figure 4. Thermoluminescence glow curve for $Sr_3Sn_2O_7:Sm^{3+}$. The sample was first exposed UV light (254 nm) and then heated from 273 K to 573 K with a rate of 1.5 K/s.

Each glow curve went through a broad maximum and fitted to approximation by at least four Gaussian curves. The four peaks located around 346 K (Peak 1), 390 K(Peak 2), 423 K (Peak 3) and 458 K (Peak 4). The activation energy *E* of those peaks, which corresponds to the depth of the traps, can be estimated by Hoogenstraaten method. The Hoogenstraaten method allow us to derive $E_1 = 0.70$, $E_2 = 0.85$, $E_3 = 1.19$, $E_4 = 1.47$ eV, which is much larger than the thermal vibration energy of 0.03 eV (350 K). This indicates that the trapped carriers should be stable at room temperature. Previously, we have proposed that an external force applied crystals induced lattice deformation, which causing local atomic displacements and generating a local electric field in the deformed region [8]. Therefore, we speculate that the local piezoelectricity is appeared in $Sr_3Sn_2O_7:Sm^{3+}$ upon compression which leads to excitation of trapped carrier, resulting in that the strong ML is occurred.

Figure 5 shows the ML and PL spectra of $Sr_3Sn_2O_7:Sm^{3+}$. The ML spectrum was in good agreement with the PL spectrum, suggesting that the ML emission originates from the same luminescence center as PL. These observed ML and PL emission spectra had five peaks at 570, 582, 610, 624, and 665 nm, which were identified as the electron transition in the Sm^{3+} ions from the excited state $^4G_{5/2}$ to the ground state ^6H_J (J = 5/2, 7/2, 9/2). This result means that ML process of $Sr_3Sn_2O_7:Sm^{3+}$ was similar to PL process of that.

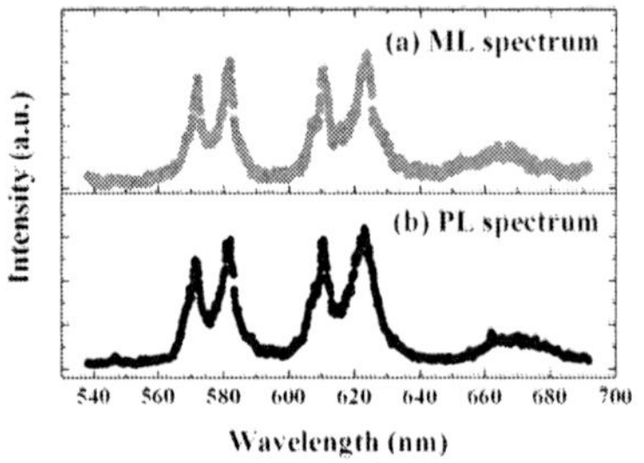

Figure 5. (a) mechanoluminescence (ML) spectrum and (b) photoluminescence (PL) spectrum from $Sr_3Sn_2O_7:Sm^{3+}$.

It is well known that the intra 4f-4f transition in trivalent rare earth ions is usually very weak because it is a parity-forbidden transition and the photon absorption/excitation is very low. However, the 4f-4f emission can be observed through the charge transfer state (CTS) [14]. As shown in Fig. 6, the PL excitation spectrum at 624 nm ($^4G_{5/2} \rightarrow {}^6H_{7/2}$) for $Sr_3Sn_2O_7:Sm^{3+}$ contain small shoulder peaks, centered at 320 nm, attributed to the CTS for O^{2-}-Sm^{3+} [15]. In addition, intense broad peaks were observed in the shorter wavelength range, which were assigned to the fundamental absorption of the host material, based on the comparison with the absorption spectra in non-doped $Sr_3Sn_2O_7$ (Fig.6). This indicated that the excitation energy was firstly absorbed by the host, and then the energy transferred to Sm^{3+} ions through a formation of CTS.

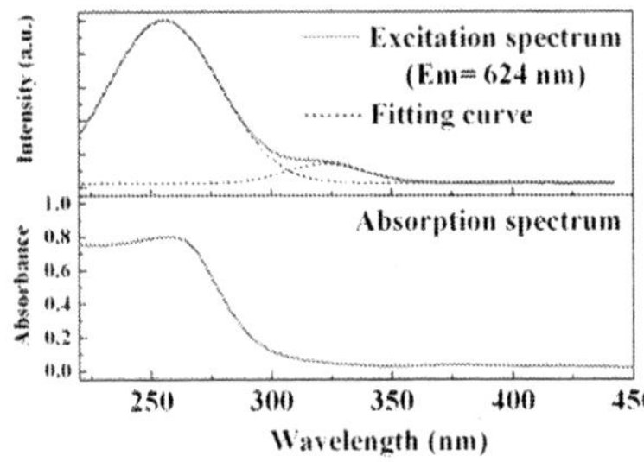

Figure 6. Excitation spectrum of $Sr_3Sn_2O_7:Sm^{3+}$ under 624 nm emission wavelength and absorption spectra of non-doped $Sr_3Sn_2O_7$.

Based on the above results, we propose the ML process for $Sr_3Sn_2O_7:Sm^{3+}$ as follows. (1) free electrons and free holes are generated in the host lattices by UV excitation. (2) the free electrons are temporarily captured by the Sm^{3+} ions, and the Sm^{3+} ions are reduced to Sm^{2+} [16, 17]. On the other hand, the free holes are trapped quasi-stably in V''_{Sr} defects. (3) When stress is applied to the material, the trapped holes are released and attracted to the O^{2-} ions around the Sm^{2+} ions. The holes are then bounded with Sm^{2+}, which leads to the formation of the Sm^{3+}-CTS [18, 19]. The non-radiative relaxation from the CTS to the $^4G_{5/2}$ excited state in the Sm^{3+} ions occurs. (4) Finally, the electron transition from $^4G_{5/2}$ to the ^6H_J (J = 5/2, 7/2, 9/2) ground state gives rise to the characteristic line emissions.

CONCLUSIONS

In conclusion, we have found the ML phenomenon in Sm^{3+}-doped $Sr_3Sn_2O_7$ perovskite-related oxides. The $Sr_3Sn_2O_7:Sm^{3+}$ emitted a strong reddish-orange light when subjected to a compressive load, which was visible to the naked eye. Therefore, $Sr_3Sn_2O_7:Sm^{3+}$ is a promising candidate for stress sensors. Furthermore, we have investigated ML properties of $Sr_3Sn_2O_7:Sm^{3+}$. The emission was identified intra-4f transition of Sm^{3+} ions based on the comparison between photoluminescence and ML spectrum. The ML intensity gradually decreased with repeating the compressive load, but it was recovered back completely by irradiating ultraviolet light (254 nm), of which phenomenon has already been known in many ML materials, indicating that ML is intimately related to the charge trap. The depth of trap of $Sr_3Sn_2O_7:Sm^{3+}$ was estimated E_1 = 0.70, E_2 = 0.85, E_3 = 1.19, E_4 = 1.47 eV from the results of the thermoluminescence measurement, indicating that trapped carrier cannot be activated at the room temperature. And excitation spectrum revealed that efficient energy transfer from the host to the Sm^{3+} ions through a formation of CTS. The present results suggest that charge traps and CTS between Sm^{3+}-O^{2-} are important roles for ML performance.

ACKNOWLEDGMENTS

The authors thank their colleagues at National Institute of Advanced Industrial Science and Technology (AIST) for assistance and advice. This work was partially supported by the JST CREST program.

REFERENCES

[1] F. S. Galasso, *Structure, Properties and Preparation of Perovskite-type compounds* (Pergamon, Oxford, 1969).
[2] K. Koumoto, Y. Wang, R. Zhang, A. Kosuga, and R. Funahashi, *Annu. Rev. Mater. Res.* **40**, 363 (2010).
[3] X. S. Wang, C. N. Xu, H. Yamada, K. Nishikubo, and X. G. Zheng, *Adv. Mater.* **17**, 1254 (2005).
[4] K. Sakai, T. Koga, Y. Imai, S. Maehara, and C. N. Xu, *Phys. Chem. Chem. Phys.* **8**, 2819-2822 (2006).
[5] C. Z. Li, Y. Imai, Y. Adachi, H. Yamada, K. Nishikubo, and C. N. Xu, *J. Am. Ceram. Soc.* **90**, 2273-5 (2007).
[6] C. N. Xu, X. G. Zheng, M. Akiyama, K. Nonaka, and T. Watanabe, *Appl. Phys. Lett.* **76**, 179 (2000).
[7] C. N. Xu, T. Watanabe, and M. Akiyama, and X. G. Zheng, *Appl. Phys. Lett.* **74**, 2414 (1999).
[8] L. Zhang, C. N. Xu, H. Yamada, and N. Bu, *J. Electrochem. Soc.* **157**, J50-J53 (2010).
[9] S. Kamimura, H. Yamada, and C. N. Xu, *J. Luminescence.* **132**, 526 (2012).
[10] S. Kamimura, H. Yamada, and C. N. Xu, *Appl. Phys. Lett.* **101**, 091113 (2012).
[11] M. A. Green, K. Prassides, P. Day, and D. A. Neumann, *Int. J. Inorg. Mater.* **2**, 35 (2000).
[12] X. H. Xu, Y. H. Wang, W. Zeng, and Y. Gong, *J. Electrochem. Soc.* **158**, J305 (2011).
[13] S. Kamimura, H. Yamada, and C. N. Xu, *Appl. Phys. Lett.* **101**, 091113 (2012).
[14] H. Yamamoto, M. Mikami, Y. Shimomura, and Y. Oguri, *J. Luminescence.* **87**, 1079 (2000).
[15] B. G. Lei, S. Q. Man, Y. L. Liu, and S. Yue, *Mater. Chem. Phys.* **124**, 912 (2010).
[16] Y. L. Huang, W. X. Zhao, L. Shi, and H. J. Seo, *J. Alloys. Comp.* **477**, 936 (2009).
[17] M. Nogami, and K. Suzuki, *Adv. Mater.* **14**, 923 (2002).
[18] C. W. Struck, and W. H. Fonger, *Phys. Rev. B.* **4**, 22 (1971).
[19] C. Pedrini, *Phys. State. Sol. a.* **202**, 185 (2005).

Mater. Res. Soc. Symp. Proc. Vol. 1492 © 2013 Materials Research Society
DOI: 10.1557/opl.2013.369

Group III-Nitride Based Electronic and Optoelectronic Integrated Circuits for Smart Lighting Applications

J. Justice, A. Kadiyala, J. Dawson and D. Korakakis
Lane Department of Computer Science and Electrical Engineering, West Virginia University, Morgantown, WV 26506 U.S.A.

ABSTRACT

With general lighting applications being responsible for over 20% of the energy consumption in the United States, advances in solid-state lighting have the potential for considerable energy and cost savings. The United States Department of Energy predicts that the increased use of solid state lighting will result in a 46% lighting consumption energy savings by the year 2030. Smart lighting systems have the potential for reducing energy costs while also providing a means for short distance data transmission via free space optics. The group III-nitride (III-N) family of materials, including aluminum nitride (AlN), gallium nitride (GaN), indium nitride (InN), their binary and ternary alloys, are uniquely situated to provide light emitting diodes (LEDs) across the full visible spectrum, photodetectors (PDs) and high power, high speed transistors. In this work, aluminum gallium nitride (AlGaN) / GaN high electron mobility transistors (HEMTs) and indium gallium nitride (InGaN) photodiodes (PDs) are fabricated and characterized. HEMTs and LEDs (or PDs) are grown on the same substrate for the purpose of creating electronic and optoelectronic integrated circuits.

INTRODUCTION

Since the successful fabrication of the first blue III-N based light emitting diodes (LEDs) by Shuji Nakamura *et al.* [1], GaN LEDs have become ubiquitous in the solid-state lighting industry, making up more than 90% of the market share for white light LEDs [2]. In recent years, III-N based high electron mobility transistors (HEMTs) have found a niche market for high-power, high-speed electronics applications [3]. Smart lighting techniques, such as free space optic (FSO) systems have also begun to be integrated into general lighting for short distance, line-of-sight communication [4-5]. This application is especially attractive for secure communication within an office building, for example.

To date, current FSO systems usually employ a solid-state lighting device that is pulse width modulated using silicon logic circuits on separate chips. The light turns on and off so fast, it is not noticed by the human eye, but can be detected by photodiodes. In this work, the authors propose to use GaN HEMTs on the same chip as LEDs and PDs to drive the emitters and amplify the detectors respectively. GaN LEDs have previously been reported [6-7], and are not the focus of this article. Instead, HEMTs and PDs are fabricated and characterized. In addition, a method for growing and fabricating HEMTs and LEDs on the same substrate is discussed.

EXPERIMENT

For this work, group III-nitride based LEDs, PDs, and HEMTs were fabricated separately on insulating GaN epilayers. The GaN epilayers were grown in an Aixtron 200/4 RF-S MOVPE system on AlN buffer layers on sapphire substrates. Trimethylgallium (TMGa), trimethlyalumium (TMAl) and ammonia (NH_3) were used as the Ga, Al and N precursors respectively. GaN epilayers were grown using 100 µmol/min of TMGa, a V/III ratio of approximately 700 and a reactor temperature of 980 °C. Hydrogen and nitrogen were used as carrier gases for all growths.

HEMT devices consisted of two layers, 25 nm of Si-doped $Al_xGa_{1-x}N$ on 375 nm of insulating GaN. $Al_xGa_{1-x}N$ films were grown with TMAl at 34 µmol/min, TMGa at 26 µmol/min and a V/III ratio of approximately 890. For $Al_xGa_{1-x}N$ films, the reactor temperature was 1000 °C.

LED and PD devices were of the same basic structure consisting of a 1.5 µm Si-doped GaN epilayer, a five period $In_xGa_{1-x}N$ /GaN multi-quantum well (MQW), a 10 nm Mg-doped $Al_xGa_{1-x}N$ electron blocking layer and a 140 nm Mg-doped GaN layer. For these growths, precursors included TMGa, TMAl, triethylgallium (TEGa), trimethylindium (TMIn), Bis(cyclopentadienyl)magnesium (Cp_2Mg) and NH_3. The MQW was grown at 695 °C using 6 µmol/min of TMIn, 21 µmol/min of TEGa, and a V/III ratio of approximately 4800 and 6200 for InGaN well layer and GaN barrier layer respectively.

For the integrated circuits, a selective area growth (SAG) method was used to grow LEDs and HEMTS on the same substrate. Growth conditions of the individually optimized devices were not modified from the above. Plasma enhanced chemical vapor deposition (PECVD) was used to deposit silicon dioxide on the GaN epilayers. They were then patterned using standard optical photolithography procedures and wet etching with buffered oxide etchant (BOE). The substrates were placed back into the reactor and HEMT structures were grown as shown in Figure 1(a). The remained SiO_2 was etched off in BOE. The sample was cleaned and another layer of SiO_2 was deposited on top of the HEMTs. The layer was patterned to remove the SiO_2 in between the HEMTs. The LEDs were grown in the unmasked areas as shown in Figure 1(b). Finally, the second SiO_2 mask was etched away in BOE. The samples were placed in an ultrasonic bath for 15 min to remove the dielectric mask from on top of the HEMT structure. The second dielectric mask was harder to remove and the authors believe a thin aluminum oxide layer forms at the SiO_2 / AlGaN interface during LED growth. Further work is needed to understand what is happening to the AlGaN layer during LED growth and its effect on transistor performance. An SEM image of the HEMT and LED structures grown on the same substrate is shown in Figure 2.

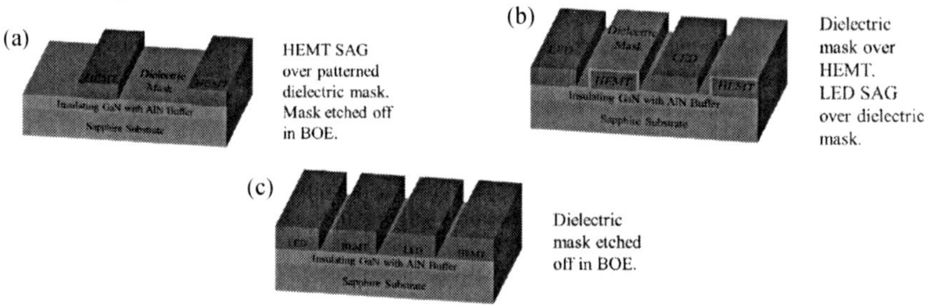

Figure 1. Growth steps for GaN based optoelectronic and electronic integrated circuits.

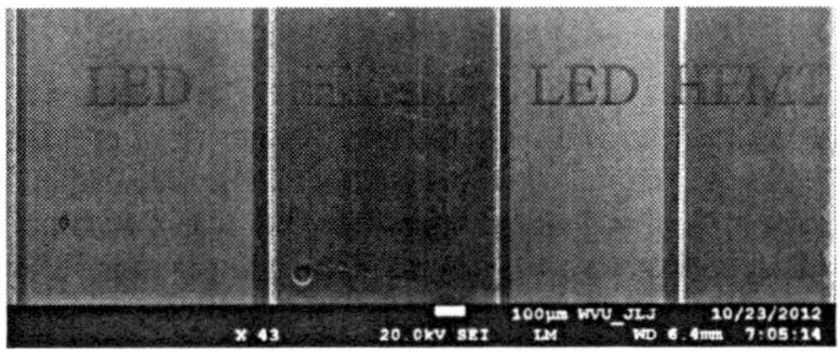

Figure 2. SEM image showing the LED and HEMT structures after growth.

DISCUSSION

HEMT and PD device characterization is discussed in the following sections. Optoelectronic and electronic integrated circuits are discussed thereafter. A more detailed discussion on LED device fabrication and characterization can be found in [6-7].

High Electron Mobility Transistors (HEMTs)

Figure 3 shows an optical image of a fabricated HEMT. In this image, the drain and source contacts are entirely on top of the AlGaN mesa. Other devices were fabricated with the Ohmic contacts overlapping the mesa edges. It was found that devices with Ohmic contacts on top of the mesa had an offset in the drain sweep. Currently this is attributed to the electron barrier in the band structure of the AlGaN/GaN heterojunction. When the contacts overlap the mesa edge, the drain sweeps pass through the origin as can be seen in Figure 4(a). This figure shows the drain sweeps from two devices with 9 and 30 µm channel lengths (labeled in Figure 3) respectively. The 30 µm and 9 µm devices have a max drain current of about 250 mA/mm and 125 mA/mm at 0 V gate bias respectively. Figure 4(b) shows the gate sweep of a 30 µm channel length device. The current through the device is reduced approximately four orders of magnitude when it is in the off state. The turn on voltage is around -3.5 V gate bias and the maximum transconductance is about 45 mS/mm at -1 V gate bias.

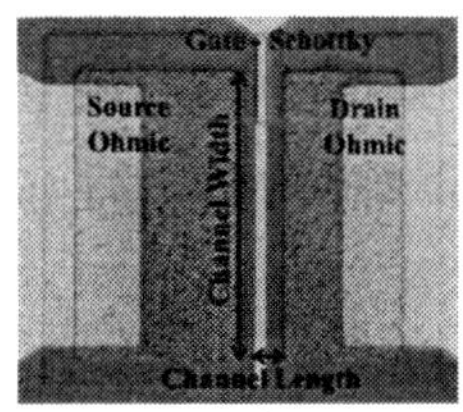

Figure 3. Optical image of a fabricated HEMT.

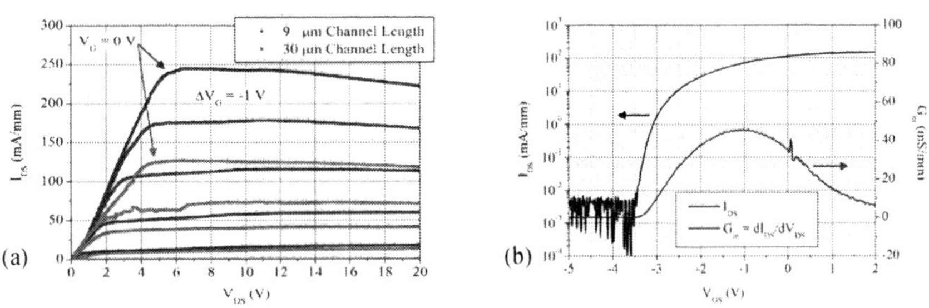

Figure 4. Drain sweep (a) and gate sweep (b) typical of fabricated HEMTs.

Photodiodes (PDs)

Figure 4(a) shows an optical image of one fabricated PD. The X-shaped contact is the p-contact and the n-contact is all the area between devices. Figure 4(b) shows the measured photocurrent when the device is illuminated with a commercially available InGaN based LED with a peak emission wavelength of 405 nm. Devices were fabricated with and without an indium tin oxide (ITO) transparent conducting layer (TCL). The measured photocurrent was ~3 times larger in devices with the ITO TCL.

(a)
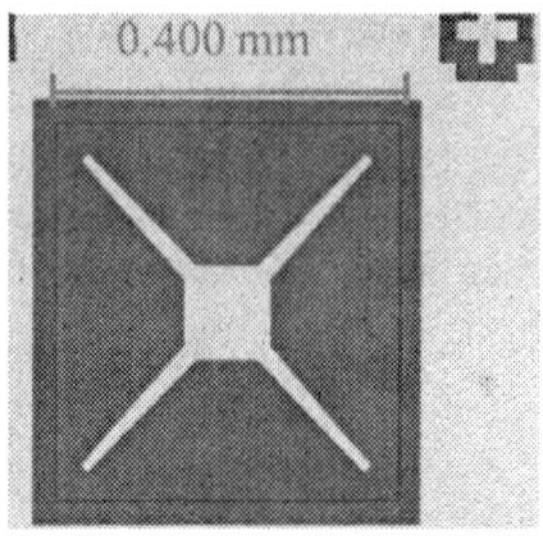

(b)
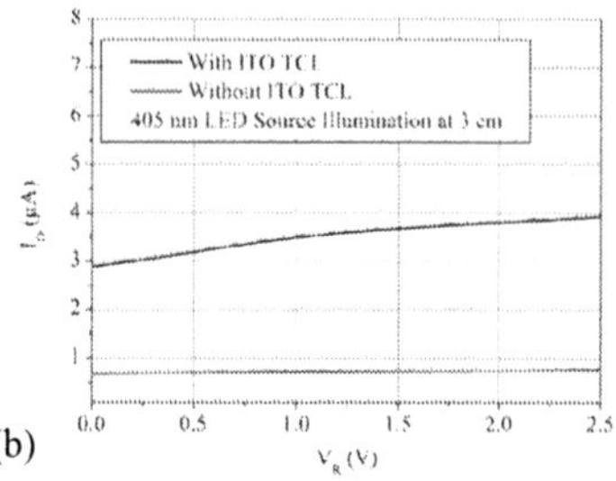

Figure 5. Fabricated PD (a) and photocurrent with 405 nm source (b).

Optoelectronic and Electronic Integrated Circuits

Figure 7 on the next page shows a process for fabricating HEMTs and LEDs on the SAG chip. For this process flow, the HEMTs must be fabricated first because the Ohmic contacts require an 800 °C anneal that would destroy the contacts of the LED. This annealing also serves to activate the Mg acceptors in the p-type GaN prior to LED fabrication. For the HEMT devices, the Ohmic and Schottky contact stacks are Ti/Al/Ni/Au (30/225/80/40 nm) and Ti/Au (80/40 nm) respectively. For LED devices the p- and n-contact stacks are Ni/Au (30/150 nm) and Ti/Al/Ti/Au (2/30/100/150 nm) respectively. The bonding pads for both devices are 600 nm Au. Future photomask design will incorporate HEMT and LED bonding pad deposition into the same fabrication step. An optical image of the devices after fabrication is shown in Figure 6.

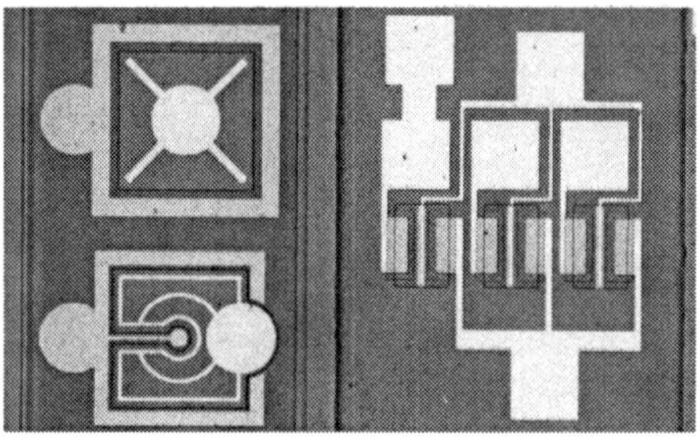

Figure 6. Optical image of LEDs/PDs (left) and HEMTs (right) fabricated on a single substrate.

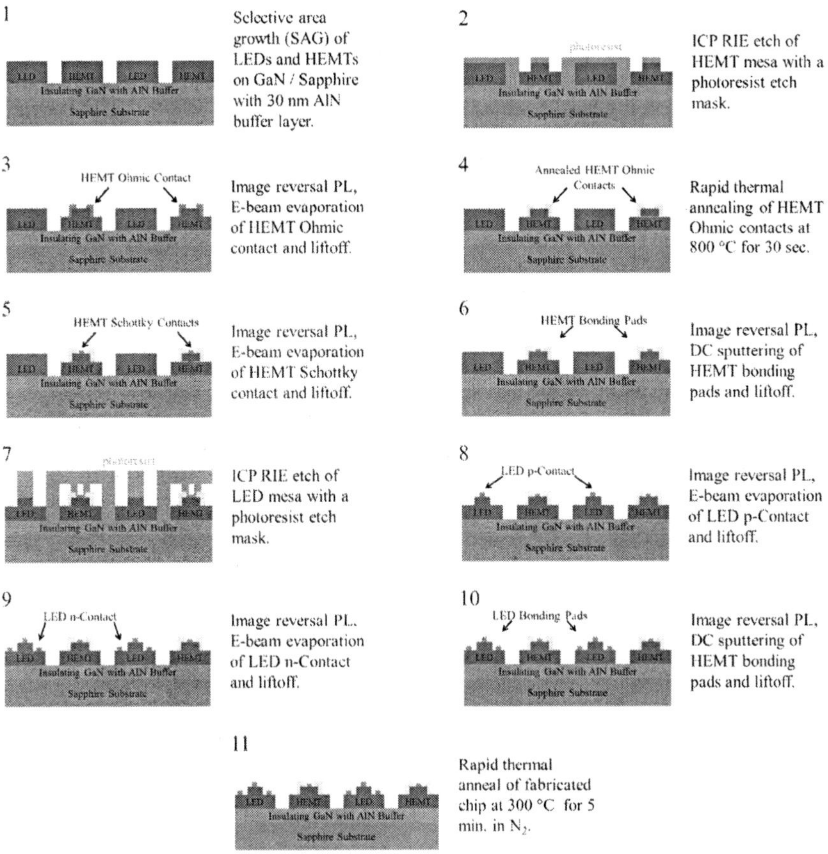

Figure 7. Process flow for the fabrication of optoelectronic and electronic devices on the same substrate.

CONCLUSIONS

LEDs were grown, fabricated and characterized in previous work [6-7]. HEMTs and PDs have been grown, fabricated and characterized independently on separate substrates. HEMTs fabricated in this work had drain currents of about 250 mA/mm and 125 mA/mm at 0 V gate bias for 30 μm and 9 μm channel length respectively. The 30 μm channel length devices had a turn on voltage around -3.5 V and maximum transconductance of approximately 45 mS/mm at -1 V gate bias. PDs fabricated in this work showed a three-fold increase in photocurrent when an ITO

TCL was deposited on top of the p-contact. A growth and fabrication process for integrating the optoelectronic and electronic devices on the same substrate is presented.

During fabrication of the integrated circuit, the SiO_2 mask was not easily removed from the HEMT structure after LED growth. It can be speculated that a thin aluminum oxide or aluminum oxynitride layer is formed at the dielectric / AlGaN interface during LED growth. The dielectric mask was successfully removed by placing the sample in BOE and then into an ultrasonic bath for 15 min. Further investigation is needed to identify what is happening at the dielectric / AlGaN interface during LED growth and what effect it has on HEMT operation.

If the dielectric stripe mask is too wide, GaN parasitic growth begins to form on top of the SiO_2 during both HEMT and LED growth. These islands cannot be removed during the wet etching step. Subsequently, devices cannot be fabricated in those areas. It was found that the ideal stripe width was 400 – 750 μm.

Surface diffusion of the group III species across the dielectric mask results in a change of the local V/III ratio at the growth site effectively changing the growth rate. The growth conditions of PDs and HEMTs have not yet been optimized for the case of the integrated circuit and the heterostructure layers are not the same thickness as they are in the independent case. As a result, the fabricated integrated devices did not behave in the same manner as the discrete devices. Further work is needed to optimize the SAG of PD and HEMT structures, however, it has been shown that devices will work when grown using this method [8].

ACKNOWLEDGEMENTS

J. Justice was supported by the Department of Defense (DoD) through the National Defense Science & Engineering Graduate Fellowship (NDSEG) Program. This work was funded in part by a grant from NASA Goddard to the West Virginia Robotic Technology Center. The authors also acknowledge use of the WVU Shared Research Facilities.

REFERENCES

[1] S. Nakamura, M. Senoh and T. Mukai, *Jpn. J. Appl. Phys.*, **32** (1993) L8.

[2] Navigant Consulting, Inc., "Energy Savings Potential of Solid-State Lighting in General Illumination Applications", Prepared for: Solid-State Lighting Program, U.S. DOE, (2012).

[3] S. N. Mohammad, A. A. Salvador and H. Morkoc, *Proc. IEEE*, **83**(10) (1995) 1306-1355.

[4] H. Willebrand and B. Ghuman, "Free Space Optics: Enabling Optical Connectivity in Today's Networks", 1st ed., (Sams, 2001).

[5] S. Deng and Z. R. Huang, Conference on Lasers and Electro-Optics (CLEO), JWA92 (2011).

[6] K. Lee, L. E. Rodak, V. Kumbham, V. Narang, A. Kadiyala, R. Goswami, B. A. Bearce, J. Peacock, K. Hite, J. M. Dawson, L. A. Hornak and D. Korakakis, *Mater. Res. Soc. Symp. Proc.*, **1396**, (2012).

[7] K. Lee, L. E. Rodak, V. Kumbham, V. Narang, J. S. Dudding, R. Rahimi, S. Kuchibhatla, L. Hornak, and D. Korakakis, *Mater. Res. Soc. Symp. Proc.*, **1288**, (2011).

[8] L. E. Rodak, K. Lee, V. Kumbham, V. Narang, L. A. Hornak, and D. Korakakis, *Mater. Res. Soc. Symp. Proc.*, **1288**, (2011).

Mater. Res. Soc. Symp. Proc. Vol. 1492 © 2013 Materials Research Society
DOI: 10.1557/opl.2013.152

Microstructure of epitaxial AlN layers on sapphire substrates deposited by physical vapor deposition

Sandeep Kohli, Boris Druz, Adrian Devasahayam, Arindom Datta, Frank Cerio

Veeco Instruments Inc., 1 Terminal Drive, Plainview NY 11803, USA

ABSTRACT

Asymmetric (10L) XRD peaks have been employed as a measure of epitaxial quality for aluminum nitride (AlN) nucleation layers (NL) deposited on sapphire substrate. Epitaxial AlN films have been deposited on 2-6" sapphire substrate by reactive sputtering. FWHM of AlN (103) and (105) were found to be an excellent indicator of quality of AlN films for GaN growth. AlN films produced nucleation layers with highly reproducible microstructure and GaN film growth. NLs had in-plane and out-of-plane texture as evident by the pole-figure results and selected area diffraction pattern. Based on electron microscopy results, AlN film thickness for complete atomic ordering was estimated to be 6-7 nm and most of the edge dislocations were seen in the first 20 nm of the film. Excellent thickness and texture uniformity were seen on planar and patterned sapphire substrates. A compressive stress of 2.9±0.2 GPa was seen in our BKM films. The maximum screw and edge dislocation densities of films were found to be ~3 x 10^8 cm^{-2} and ~9 x 10^9 cm^{-2} respectively. The root mean square roughnesses of A-polar films were found to be < 0.3 nm.

INTRODUCTION

Since the earliest work by Nakamura etal[1] Gallium Nitride (GaN) based solid state lighting (SSL) has matured enough to enable commercialization of this technology for large scale application in consumer electronics as well as recent entry in the area of house-hold lighting[2]. While GaN growth on sapphire and silicon carbide [3]has been the mainstay of the technology, GaN on silicon (Si) has made some recent news for SSL [4]. However, GaN on Si is yet to mature enough for large scale consumer applications.

Efforts are being continuously made to improve status of current GaN based SSL on sapphire. Amongst the methods being explored, presence of aluminum nitride (AlN) nucleation layer[5, -8] and use of patterned sapphire substrates[9] are touted as some major improvements in the technology to improve, the external as well as internal quantum efficiencies of GaN based SSL technology on sapphire leading to lower costs and more affordability for mainstream house-hold lighting. Some recent results have reported significant results in this direction [10]. Also, presence of a thin seed AlN layer also eliminates 1st step of the growth of low temperature GaN layer (~2.5 um) that is frequently encountered in standard 2-step nucleation process (MOCVD process for LEDs grown on sapphire). Although, AlN may be deposited easily using metalorganic chemical vapor deposition (MOCVD) methods, the slow rate of deposition using MOCVD is a limiting factor. Hence, physical vapor deposition (PVD) techniques like reactive sputtering that enable comparatively faster growth rates and low cost of ownership may be preferred. However, the literature in this direction has been limited in scope [11, 12]. Most of the work involving AlN nucleation layer reported has been done with MOCVD processes [13].

In the present article we report our results for results for reactively sputtered AlN films using an aluminum target using a commercially developed deposition chamber. Results indicate that high-quality AlN/ sapphire templates may be successfully fabricated and used for improved GaN growth. In the results reported in the literature, the FWHM of the rocking curves for symmetric AlN (002) and AlN (102) are typically presented as a qualitative marker for AlN film [13, 14]. AlN (102) peak is typically accessed by tilting the stage that may add to the defocusing of the X-ray diffracted beam. The focus of this work is to establish utilization of asymmetric (105) and (103) peaks as quality marker for the growth of high-quality AlN films on sapphire substrates. All the depositions were carried out in production quality deposition tools.

EXPERIMENTAL

Al-N films were reactively sputter deposited on c-plane 2"-6" sapphire substrates using aluminum target. As received, sapphire substrates without any *ex-situ* or *in-situ* surface cleaning were used in this work. The base pressure was better than 10^{-8} torr. Our standard deposition conditions involved an Ar/N_2 gas flow with the rotating substrate held at 900 °C. Prior to the start of deposition, substrate temperature was allowed to stabilize for less than 30 minutes.

Films were characterized using Panalytical double axis X-ray diffraction (XRD) system equipped with a 2-bounce monochromator and Copper X-ray source. X-ray reflectivity (XRR) measurements were also performed in the theta-2Theta range 0.2-5° with a step size of 0.005°. Spectroscopic Ellipsometry (SE) measurements were performed using J.A. Woollam spectroscopic ellipsometer at variable angle and wavelengths. SE results were modeled using WVASE software provided by the manufacturer. Scanning Electron microscopy (SEM) was performed using JEOL JSM 6500 microscope while Atomic Force Microscope (AFM) measurements were performed using Dimension 3100 AFM instrument. Electron microscopy measurements were performed using FEI Tecnai G2 F20 instrument.

RESULTS AND DISCUSSIONS

50 nm AlN films were deposited on a 2"-6" c-plane sapphire substrates held at 900 °C. Although, Al-N films with lower thicknesses (25 nm and 35 nm) were also tried, it was discovered that it is hard to control the microstructure of the thinner films leading to lot of variability in the deposited microstructure and thus film quality. Thicker films did not add much to the quality of subsequent GaN films. Thus 50 nm film thicknesses were chosen to get optimum quality and reproducibility of the film quality. A temperature of 900 °C was chosen, since at this temperature we were able to get smooth high quality textured films. Lower temperatures led to the poor crystallinity of the films while higher temperatures led to the increased surface roughness. Therefore, a substrate temperature of 900 °C was a good compromise between the crystalline quality of films and surface roughness. Partial pressure of argon and nitrogen were also optimized for best growth conditions.

The full width half maxima (FWHM) of the rocking curves for (002), (105) and (103) peaks were found to be ~ 350s, 1100s and 1200s respectively. Thicknesses of the films for each run were confirmed by X-ray reflectivity. While, FWHM of AlN (002) peak is more sensitive to screw dislocations, edge dislocations have more impact on the FWHM of peaks from planes that are not parallel to AlN (002); higher the angle of inclination of plane with respect to AlN (002), more sensitive it is towards dislocations. As it may be appreciated, AlN (105) and AlN (103) planes have different inclination angle with respect to AlN (002) planes. Hence, monitoring the FWHM of these symmetric and asymmetric peaks helped us not only in identifying the optimum conditions for growth, but also sort out samples that would not show immediate GaN recovery on the templates. In our case (50 nm films), it was found that FWHM of AlN (002), AlN (105) and AlN (103) had to be better than 380s, 1150s and 1250 s respectively; all three conditions had to be satisfied; for GaN growth.

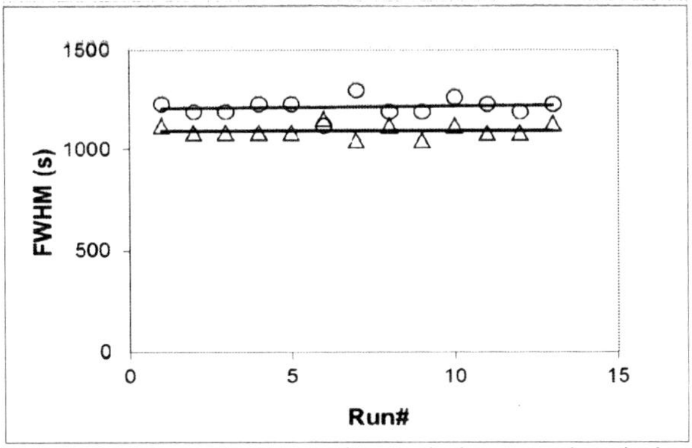

Figure 1. Repeatability run for 50 nm AlN reactively sputter deposited on c-plane Sapphire substrate. FWHM of rocking curve for AlN [O] (103) and [Δ] (105) peak are shown

Figure 1 shows the repeatability data for FWHM of rocking curves for AlN films for a typical run. Based on abovementioned results the FWHM of rocking curves for (105) and (103) could be used to effectively sort out good AlN templates *(templates on which high quality GaN growth would occur with full recovery)* from the bad samples (*templates on which poor quality GaN growth would occur with partial or no recovery*). Based on the FWHM of the rocking curves, the density of screw dislocations in AlN films were found to be ~3x 10^8 cm^2 while the density of edge dislocation (AlN (103) peak) was found to be ~ 8 x 10^9 cm^2.

The thickness uniformity of AlN films across 2-6" wafer was established using Spectroscopic ellipsometry. In determining the optimum optical constants and model, initially optical constants of sapphire substrate were modeled using Cauchy parameters. Subsequently, for the case of AlN, a Cauchy layer was used to fit the experimental results. Uniaxial anisotropy was added in optical constants to get the best fit results in the case of AlN layer. The surface roughnesses of the AlN films were measured using AFM. The root mean square (RMS) roughnesses of the films were found to be < 0.3 nm over a 5 μm x 5 μm area and 20 μm x 20 μm

(Fig 2). Based on AFM results, the layers were smooth and devoid of any defects on the layer surface for AlN (Fig 2) epilayer.

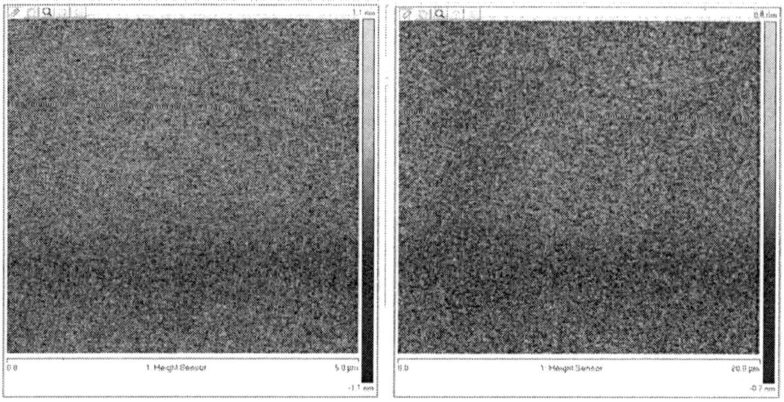

Figure 2. AFM images of 50 nm AlN on c-plane sapphire substrate

In order to confirm the epitaxial growth of AlN films, transmission electron microscopy measurements were performed. Fig 3a shows the selected area diffraction pattern of 50 nm AlN deposited on sapphire substrate. Clearly, our AlN films are single crystal and epitaxial to c-plane Al_2O_3. These results are also consistent with our pole figure results on AlN (002), (102) and (011) planes (not shown here). A lattice mismatch of ~11% was measured between AlN and Al_2O_3. In the TEM images analyzed, the AlN (0002) fringes run through the image. No obvious grain boundary could be identified. Morié fringes were used to estimate the thickness so that AlN atomic ordering is fully recovered. This thickness was estimated to be 6-7 nm in different images. Most of the edge dislocations were seen in the first 20 nm of the film. This is consistent with our observations that AlN film thicknesses below 25 nm were not suitable for GaN growth.

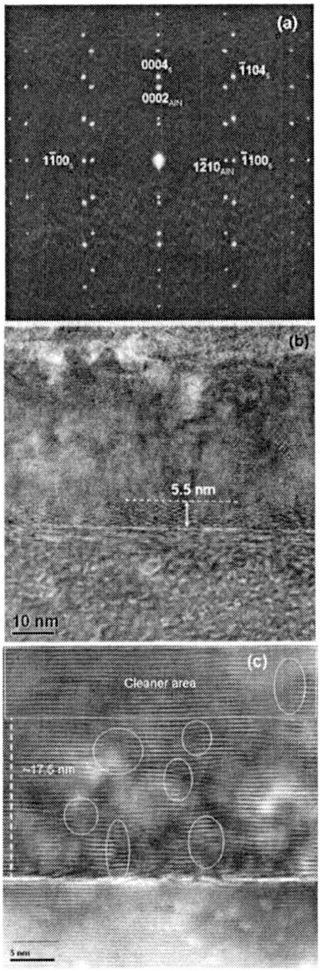

Figure 3. (a) Select area electron diffraction pattern for 50 nm AlN epitaxial layers reactively sputtered on c-plane sapphire substrate. Miller indices for some important reflections for epilayer and substrate are shown. (b) Electron micrograph of 50 nm AlN layer showing that (b) AlN atomic ordering is fully recovered in 6-7 nm and (c) most of the dislocations are see in < 20 nm of film thickness from the substrate.

CONCLUSIONS

Epitaxial AlN films have been deposited on sapphire substrate by reactive sputtering. FWHM of AlN (103) and (105) were found to be an excellent indicator of epitaxial quality of AlN films. It was observed that the FWHM of (002), (103) and (105) had to be lower than 380s, 1250s and 1150s respectively for high quality GaN growth. AlN film thickness for complete atomic ordering was estimated to be 6-7 nm. Most of the edge dislocations were seen in the first 20 nm of the film.

ACKNOWLEDGEMENT

Authors would like to acknowledge the support by Dong Lee, Hongwei Lee and Frank Lu, Veeco Instruments, Somerset, NJ for their support in GaN deposition and measurements. Financial support provided by Department of Energy (DOE) (grant # EE0005064) is also acknowledged.

REFERENCES

1. S. Nakamura, T. Mukair & M.Senoh (1991). Jap. J. Appl. Phys., 30, L1998.
2. S.E. Bennett (2010). Mat Sci Tech, 26, 1017-1028.
3. P. Gibart (2004). Rep. Prog. Phys, 67 , 667 .
4. D. Zhu, C. McAleese, M. Häberlen, C. Salcianu, T. Thrush, M. Kappers, A. Phillips, P. Lane, M. Kane, D. Wallis, T. Martin, M. Astles, N. Hylton, P. Dawson & C.Humphreys (2011). Journal of applied physics, 109, 014502(6).
5. K. Dovidenko, S. Oktyabrsky, J. Narayan & M.Razeghi (1996). J. Appl. Phys, 79, 2439-2445.
6. A. Volkova, V. Ivantsov & L.Leung (2011). J Crys Growth, 314, 113.
7. S. Yoshida, S. Misawa & S.Gonda (1983). Appl Phys lett, 42, 427.
8. H. Amano, N. Sawaki & I.Akasaki (1986). Appl. Phys. Lett., 48, 353.
9. T.S. Kim, S. Kim, Y.H. Jang & G.Y.□.Junga (2007). Appl Phys letters, 91, 171114.
10. J. Bläsing, A. Krost, J. Hertkorn, F. Scholz, L. Kirste, A. Chuvilin & U.Kaiser (2009). J Appl. Phys, 105, 033504(9).
11. C. Duquenne, M.A. Djouadi, P.Y. Tessier, P.Y. Jouan, M.P. Besland, C. Brylinski, R. Aubry & S.Delage (2008). Appl Phys Let, 93, 052905.
12. E. Valcheva, T. Paskova, S. Tungasmita, P.O.A. Persson, J. Birch, E.B. Svedberg, L. Hultman & B.Monemar (2000). Appl Phys Lett, 76, 1860.
13. Y.A. Xi, K.X. Chen, F. Mont, J.K. Kim, C. Wetzel, E.F. Schubert, W. Liu, X. Li & J.A.Smart (2006). Appl. Phys. Lett., 89, 103106 (3).
14. B.N. Pantha, R. Dahal, M.L. Nakarmi, N. Nepal, J. Li, J.Y. Lin, H.X. Jiang, Q.S. Paduano & D.Weyburne (2007). Appl Phys Lett, 90, 241101 (3).

Environment and Health

Mater. Res. Soc. Symp. Proc. Vol. 1492 © 2013 Materials Research Society
DOI: 10.1557/opl.2013.154

Characterization of carbon dioxide separation membrane with polycation nano-layers

Tatsuya Funaoka, Yusuke Daiko, Atsushi Mineshige, Tetsuo Yazawa
Department of materials science and chemistry, University of Hyogo, 2167 Shosha, Himeji, Hyogo 671-2280

ABSTRACT

Ultrathin layers of positively charged poly(diallyl dimethylammonium) choloride (PDDA) and negatively charged poly(sodium 4-styrenesulfonate) (PSS) were deposited on SiO_2/ polyethylene glycol hybrid membranes via layer-by-layer assembly technique, and carbon dioxide absorption/separation properties were investigated. Quartz crystal microbalance (QCM) measurements revealed that both PDDA and PSS nanocoatings have a good affinity for CO_2 absorption. PDDA-deposited film shows about two times higher CO_2 ideal gas selectivity compared with unmodified silica film.

INTRODUCTION

The global warming caused by carbon dioxide emission is becoming an urgent environmental problem. The removal and separation of carbon dioxide from flue gas have been considered as a promising technology for green house gas emission control. Aqueous solutions of amine-based absorbent are used for capturing carbon dioxide. However, this absorption process is discontinuous and high in cost. Compared with the absorption method, membrane separation technique has advantages such as continuous process (rapid separation) and low operation cost. Thus membrane separation has attracted much attention for practical carbon dioxide separation [1]. Recently, active research has been carried out on inorganic-organic hybrids for carbon dioxide separation prepared by sol-gel method [2,3].

For practical application, membranes with high permeance and high selectivity for carbon dioxide are required. However, there is a trade-off relationship between gas permeance and selectivity; a membrane with high permeance shows low selectivity. In this study, we focused on layer-by-layer (LBL) self-assembly technique. LBL assembly has been widely used to fabricate various type of multilayered structures [4-7]. The LBL assembly offers an easy and inexpensive solution-process and allows a variety of materials to be incorporated within the film structures. As an outline of the technique, LBL assembly is mainly conducted through electrostatic interaction; oppositely charged materials such as polyelectrolytes are alternately deposited on a charged-substrate, and an insoluble polymer complex with a few nanometer thicknesses can be obtained at room temperature and ambient atmosphere. Considering high gas permeance, film thickness is a crucial aspect. We anticipated that such a positively or negatively charged ultrathin layers effectively improve the affinity for CO_2 without a serious decrease of permeance because CO_2 is also a polarizable molecule as $O^{\delta-} = C^{\delta+} = O^{\delta-}$. The zeta-potentials of polyelectrolyte such as positively charged poly(diallyl dimethylammonium) choloride (PDDA) and negatively charged poly(sodium 4-styrenesulfonate) (PSS) are reached to approximately +50 and -50 mV, respectively [8].

In this study, PDDA and PSS layers were deposited on a CO_2 separation layer via LBL technique. Single gas permeance (He, N_2, CO_2) and selectivity for CO_2 gas are discussed in relation to the absorption of CO_2 gas.

EXPERIMENTAL

A porous alumina tube (NGK Insulator Ltd.) with pore diameter of 0.1 μm, outer tube diameter of 10 mm, inner diameter of 7 mm and length of about 7 cm was used as a substrate.

First, colloidal silica layer was deposited on the support in order to fill its large pore with silica. Ethanol and commercially available colloidal silica (Snowtex O; particle size of 10-20 nm, Nissan Chemical Industriesm Ltd.) were mixed (5 : 1 in weight ratio) and then coated on outer surface of the support by dip-coating method at withdrawn rate of 1 mm s^{-1}. This dipping process was repeated for three times, and then heat-treated at 673 K for 2 h. The dip-coating and heating cycle was repeated twice. This colloidal silica layer is called as the intermediate layer in Fig. 1.

Second, CO_2 separation layer was deposited on the intermediate layer. The sol solution was prepared as follow. TEOS was hydrolyzed in HNO_3, EtOH and PEG solution for 3 h at room temperature. $Si(OC_2H_5)_4$ (TEOS) : HNO_3 : H_2O : EtOH (1 : 0.01 : 2 : 20 mol%) and polyethylene glycol (PEG) were used. The amount of PEG was fixed to be 40 weight % for SiO_2 prepared from TEOS. After the hydrolysis of TEOS, the intermediate layer-deposited substrate was immersed into the sol solution for three times. After dip-coating the separation layer, sample was heated at 423 K for 2 h. Similar to the intermediate layer, the coating and heating cycle was repeated twice.

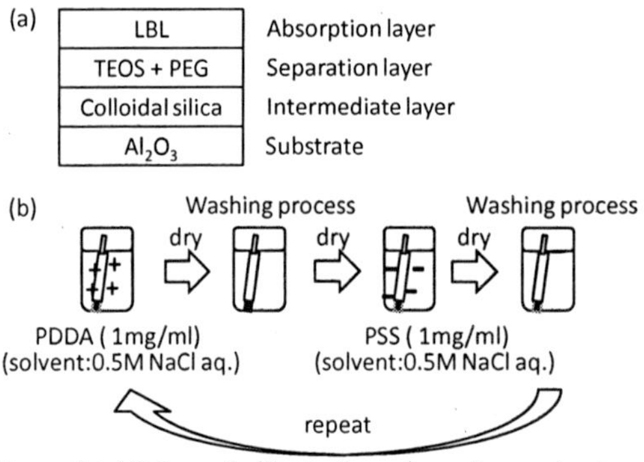

Figure 1 (a) Schematic illustration of membrane structure and (b) LBL procedures.

Finally, polycation (poly[diallyl dimethylammonium]choloride; PDDA, Mw = 100000-200000, Aldrich, 1 mg/ml of 0.5 M NaCl solution) and polyanion (poly[sodium 4-styrenesulfonate]; PSS, Mw = 70000, Aldrich, 1 mg/ml of 0.5 M NaCl solution) were deposited via layer-by layer (LBL) method. The separation-layer-coated support was immersed into PDDA solution for 7 min. The membrane was then dried for 1h under a reduced pressure, followed by rinsing with ion-exchange water and drying at 333 K. The PDDA-deposited substrate was then immersed into PSS following the same procedures as for deposition of PDDA. These LBL layers are called as the absorption layer.

Before gas permeance measurement, the obtained sample was sealed with epoxy resin. Single gas permeation through these samples at different temperatures was measured using helium, nitrogen and carbon dioxide gasses. The CO_2 ideal gas selectivity was calculated from the obtained permeance, P_{CO2}/P_{N2}, where P_{CO2} and P_{N2} are gas permeance for CO_2 and N_2, respectively.

The amount of absorbed CO_2 gas on the separation and absorption layers was monitored by using a quartz crystal microbalance (QCM) (SEIKO EG&G QCA 922) [9-12]. The QCM electrode was cleaned with a piranha solution (98% H_2SO_4 : 30% H_2O_2 = 3:1, v/v), rinsed with pure water, and dried with air flow. The electrode was alternately immersed in a PDDA and then in a PSS solution. After each absorption step, the electrode was washed and dried with air flow and then weighed in an ambient atmosphere at room temperature.

RESULT AND DISCUSSION

Gas permeation

Pore diameter of the samples after dip-coating the intermediate layer was measured using nitrogen gas absorption apparatus (BEL JAPAN INC. BEL-SORP-mini). Pore diameter was calculated to be 2 nm from the desorption branch. It was reported that SiO_2 and PEG hybrid coatings have CO_2 separation ability [13]. Figure 2 (a), (b) and (c) show the gas permeance as a function of temperature for samples without LBL coating (naked) and with PDDA, PDDA/PSS and PDDA/PSS/PDDA layers. The thermal vibration of molecular chain increases with increasing temperature, resulting into increase free volume. As the results, permeances for all the gasses increase with increasing temperature.

He gas permeance is shown in Fig. 2 (a). Compared with the naked sample, the permeance slightly decreases after the deposition of PDDA. On the other hand, a significant decrease was observed for PDDA/PSS multilayer sample. The multilayer became denser as compared with PDDA single one, and the PDDA/PSS layer blocks He gas permeation. For rapid separation, a high permeance should be required.

The molecular sizes of N_2 and CO_2 are the same as 0.32 nm (He: 0.27 nm). Since about 80% of ambient atmosphere is N_2, the ratio of P_{CO2}/P_{N2} is a crucial parameter. N_2 and CO_2 gas permeances are shown in Fig. 2 (b) and (c), respectively. Compared with the permeance for He gas, a large decrease in N_2 permeance was observed for all the samples, especially after LBL deposition. In Fig. 2 (c) about CO_2 gas permeance, similar to the results of He and N_2 gases, a relatively high permeance was observed for both naked and PDDA-coated samples as compared with PDDA/PSS and PDDA/PSS/PDDA multilayers-deposited ones.

From the results of N_2 and CO_2 gas permeances, CO_2 ideal gas selectivity was calculated and the data are shown in Fig. 2 (d). Note that PDDA-coated sample shows higher CO_2

selectivity as compared with unmodified sample (naked). Especially, at 80 °C, about two-fold increase in the selectivity was observed after deposition of PDDA. Similarly, multilayers of PDDA/PSS and PDDA/PSS/PDDA coated sample also show good CO_2 selectivity, although the CO_2 gas permeance is relatively low (approximately 1 GPU).

According to the solution-diffusion mechanism, the permeability coefficient is proportional to the product of diffusivity (D) and solubility (K);

$$P \equiv D \cdot K \qquad (1)$$

Here, solubility is correlated with the amount of absorbed CO_2 molecules on the polyelectrolyte layers. These results strongly suggest that the surface polycation layers work as a molecular antenna and effectively increase the affinity for CO_2.

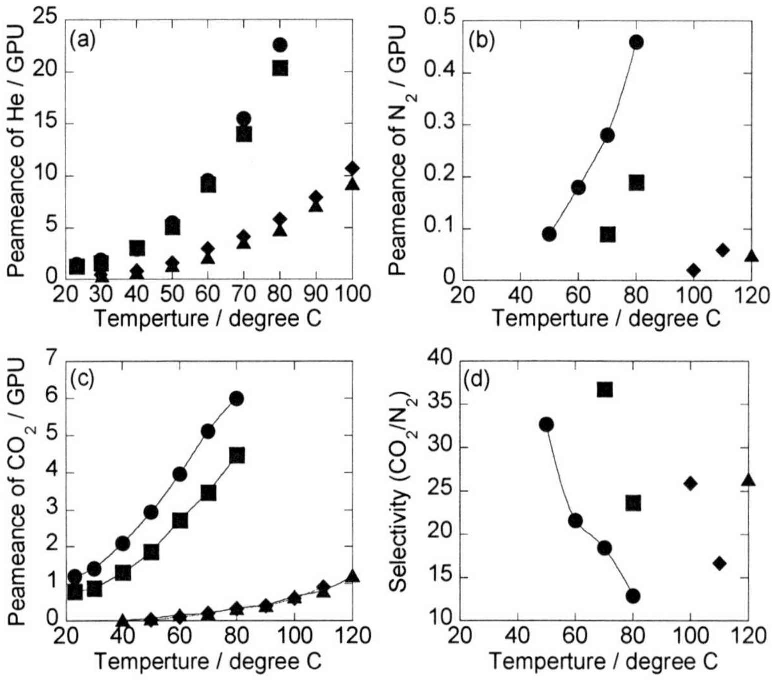

Figure 2. Gas permeance for (a) He, (b) N_2 and (c) CO_2 as a function of temperature for unmodified silica/PEG separation sample (Naked) (•), and after deposition of PDDA (■),

PDDA/PSS(♦) and (PDDA/PSS/PDDA) (▲). (d) CO_2 ideal gas selectivity (P_{CO2}/P_{N2}) as a function ot temperature. [GPU = 10^{-6} [cm^3(STP)cm/(cm^2 s cmHg)].

QCM measurement.

To ensure the absorption of CO_2 on the PDDA or PSS layers, QCM measurements were carried out. Bare QCM electrode was also used as a standard to check the accuracy (noise) of QCM frequency. QCM electrode was placed in a glass-bottle and CO_2 was flowed during the measurement at the rate of 10.57 mL/min. Figure 3 shows the changes in QCM frequency with CO_2 exposure time. In Fig. 3, $\Delta F < 0$ means the weight increase of the electrode. No significant decrease was observed for bare QCM electrode. On the other hand, it was immediately apparent that ΔF decreases clearly after the deposition of PDDA. Further decrease in ΔF was observed for the PSS-deposited sample. These QCM results revealed that there is an attraction force between the charged surface layer (both PDDA and PSS) and CO_2 molecule. In the case of PSS deposited sample, the interaction between $-SO_3^-$ groups of PSS and CO_2 ($-SO_3^- \cdots C^{\delta+}O_2$) may be too high compared with that between $-N^+$ of PDDA and CO_2. Further quantitative analysis about the amount of absorbed CO_2, thickness/density of layers, permeance and those relations to the structure of various polyelectrolytes are in progress.

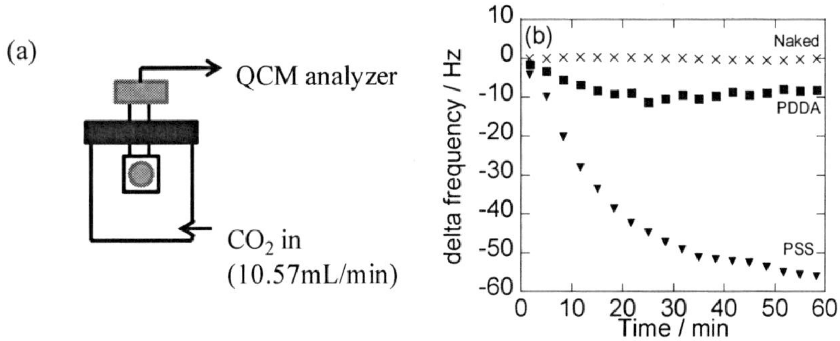

Figure 3 (a) Schematic illustration of QCM measurement, (b) Changes in QCM frequency with CO_2 exposure time for no polyelectrolyte (bare QCM electrode) (•), PDDA-deposited (■) and PSS-deposited samples (▲).

CONCLUSIONS

Ultrathin layers of positively charged poly(diallyl dimethylammonium)choloride (PDDA) and a negatively charged poly(sodium 4-styrenesulfonate) (PSS) were deposited on silica/PEG hybrid membrane, and permeance and selectivity of CO_2 gas were investigated. QCM results revealed both the PDDA and PSS have an affinity for CO_2 likewise molecular antenna, and these charged ultrathin layers absorb CO_2 effectively. The CO_2 ideal gas selectivity of PDDA-deposited sample was about two times higher than that of unmodified sample. The nanolayers prepared via LBL assembly are a promising candidate for improving CO_2 gas separation in temperatures about 25 up to 100°C.

REFERENCES

1. Feasibility paper on new energy technologies related to global environment problems NEDO-P-8914(NEDO, Tokyo, 1990) p. 84.
2. K. Kuraoka, H. Tanaka, T. Yazawa, *J. Mat. Sci. Lett.* **15**,1.(1996).
3. K. Kuraoka, N. Kubo, T. Yazawa, *J.Sol-Gel Sci Technol.* **19**, 515.(2000).
4. K. Ariga ,J. P. Hill, Q. Ji, *Phys. Chem. Chem. Phys.* **9**, 2319 (2007).
5. G. Decher and J. Schlenoff eds., *Multilayer Thin Films. Sequential Assembly of Nanocomposite Materials*, Wiley-VCH Weiheim GmbH (2003).
6. Decher, G. Hong, J.D. Makromol. *Chem. Macromol. Symp.* **46**, 321(1991) .
7. Y. Li.; F Liu, *J. Sun. Chem. Commun.* **45**, 2730.(2009).
8. K. Katagiri , F. Caruso (2005) *Adv. Mater.* **17**, 738.(2005).
9. B.A. Martin, H. E. Hager, *J. Appl. Phys.* **65**, 2630.(1989).
10. B.A. Martin, H. E. Hager, *J. Appl. Phys.* **65**, 2627.(1989).
11. T. Viitala, J.T. Hautala, J. Vuorinen, *Langmuir.* **23**, 609.(2007).
12. Stephen J. Martin, Victoria Edwards Granstaff, Gregory C. Frye. *Anal. Chem.* **63**, 2272.(1991).
13. Kim H, Lim C, Hong S, *J. Sol-gel Sci. Techonol.* **36**, 213 (2005).

Mater. Res. Soc. Symp. Proc. Vol. 1492 © 2013 Materials Research Society
DOI: 10.1557/opl.2013.220

Photocatalytic Properties of TiO_2 / WO_3 / FTO Multi-layer Structures Prepared by Spray Pyrolysis Deposition

Masahiko Maeda and Takahiro Horikawa
Kanazawa Institute of Technology, 7-1 Ohgigaoka, Nonoichi, Ishikawa, 921-8501, Japan

ABSTRACT

Visible-light photocatalytic property of TiO_2 / WO_3 / FTO multi-layer structure was investigated. In this structure, the visible-light absorption takes place in the WO_3 layer, and generated electrons and holes diffuse into the FTO and TiO_2 layers, respectively. The holes diffused into the valence band of the TiO_2 layer produce the hydroxyl radicals due to oxidation of H_2O and/or OH^- adsorbed on the TiO_2 surface. The FTO bottom layer, which acts as a scavenger of electrons, contributes to an effective charge separation. As a result, the visible-light photocatalysis is enhanced in this structure.

INTRODUCTION

In recent years, titanium dioxide (TiO_2) with super-hydrophilic and photocatalytic characteristics has received a great deal of attention. In environmental fields, particularly, such materials are important because of their ability to detoxify environmental pollutants [1]. When the TiO_2 is irradiated with ultraviolet-light (UV-light), organic compounds are decomposed on their surface and water spreads evenly to their super-hydrophilic surface, as a result, surface self-cleaning can be easily realized. Because the bandgap energy of TiO_2 is 3.2 eV, UV-light of wavelength of 380 nm or less is necessary to fulfill their photocatalytic properties. Even in outdoors during daytime, the intensity of UV-light is only about 1 mW/cm^2. Under these circumstances, the demand for the photocatalytic materials that can respond to visible-light has strengthened day by day. For development of the visible-light response TiO_2 based photocatalytic materials, anion doped TiO_2 has been intensively investigated until now [2]. The other approach has been reported that bilayer structures of TiO_2 and other metal oxide films with bandgap less than 3 eV, for example WO_3, are effective [3]. In these structures, however, photocatalytic activity is relatively low because of insufficient spatial charge separation.

In this study, we investigated on the visible-light response of photocatalysis for the TiO_2 / WO_3 / FTO multi-layer structures prepared by spray pyrolysis deposition (SPD). In these structures, the photocatalytic activity is considerably enhanced using conductive FTO underlayer. Comparing with the structures inserted SiO_2 layer, the carrier transfer from WO_3 to TiO_2 and FTO layers takes place and consequently spatial charge separation is caused sufficiently.

EXPERIMENT

Film preparation

Figure 1 shows schematic illustration of SPD apparatus which is equipped with a spraying nozzle, a hot plate, a compressor, and peristaltic pump. Reagent grade Titanium (IV) acetylacetonate, Tungsten (VI) oxide, and Tin (IV) chloride pentahydrate were used as starting chemicals. The precursor solutions for TiO_2 and SnO_2 films were 0.1 mol/L ethanol solution of $[Ti(C_5H_7O_2)_2]$ and 0.2 mol/L ethanol solution of $SnCl_4 \cdot 5H_2O$, respectively. For FTO film, NH_4F solution was mixed in $SnCl_4$ ethanol solution. The precursor $(NH_4)_2WO_4$ for WO_3 film was obtained by mixing WO_3 powder with ammonium solution. These precursor solutions were atomized by a spraying nozzle at an air pressure of 0.15 MPa. The droplets were transported onto a heated Si or quartz substrate by spraying air. The films were deposited in order of SnO_2 (or FTO), WO_3, and TiO_2. Substrate temperature and spraying time were 350 °C and 2.5 min for SnO_2 and TiO_2, and 170 °C and 2 min for WO_3, respectively. Each film was annealed at 500 °C for 4 hours prior to continuous spraying. The experimental parameters were listed in Table 1.

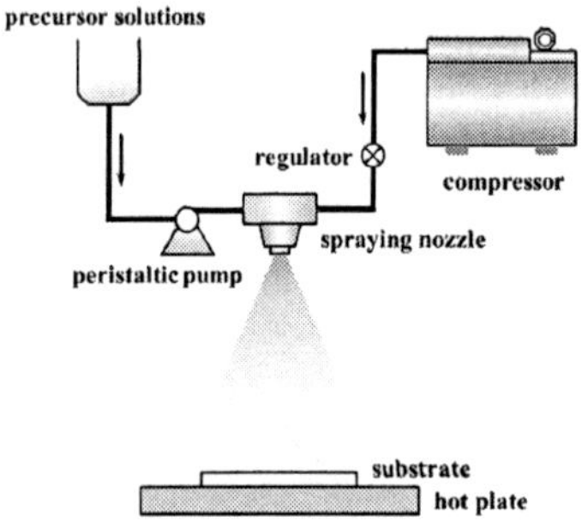

Figure1. Schematic illustration of SPD apparatus.

Table I. Experimental parameters.

Starting materials	$Ti(C_5H_7O_2)_2$
	WO_3 powder
	$SnCl_4 \cdot 5H_2O$
Film thickness	500 nm
Flow rate of solution	15 mL / s
Working distance	50 cm
Spraying time	2 ~ 2.5 min
Substrate temperature	170 ~ 350 °C

Film evaluation

Surface morphology, crystallinity and transmittance of the films were evaluated by scanning electron microscopy (SEM), X-ray diffraction (XRD), UV-VIS spectroscopy, respectively.

Photocatalytic activities of the films were evaluated by pigment degradation measurement. The light source of visible-light used was 0.5 mW/cm^2 halogen lamp whose UV fragment was cut by UV filter. The degradation rate of methyleneblue as a pigment was evaluated by measuring the changes in absorbance of 654 nm using UV-VIS spectrometer.

RESULTS and DISCUSSION

Film characterization

Figures 2 (a) and (b) show the top and cross-sectional views of the TiO_2 / WO_3 / FTO multi-layer structure, respectively. The SEM photographs show obvious three-layer structure with smooth surface. Each film thickness was about 500 nm.

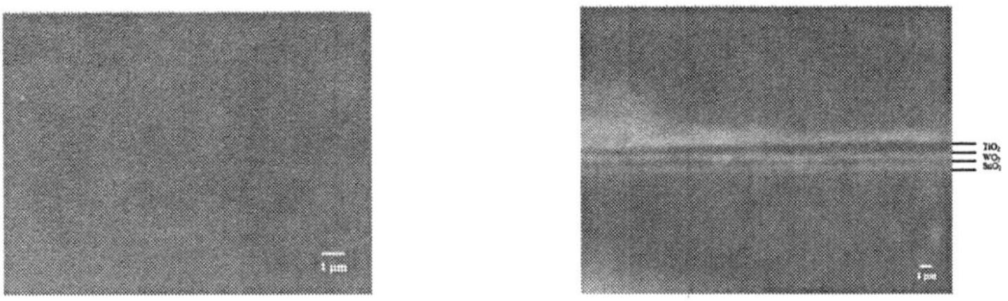

(a) (b)

Figure 2. Top (a) and cross-sectional (b) views of TiO_2 / WO_3 / FTO multi-layer structure.

Figure 3 shows X-ray diffraction patterns of the TiO_2 film and TiO_2 / WO_3 / FTO multi-layer structure. The diffraction peaks from each film were observed in the TiO_2 / WO_3 / FTO multi-layer structure, and it is noticed that the TiO_2 top-layer has anatase crystallinity.

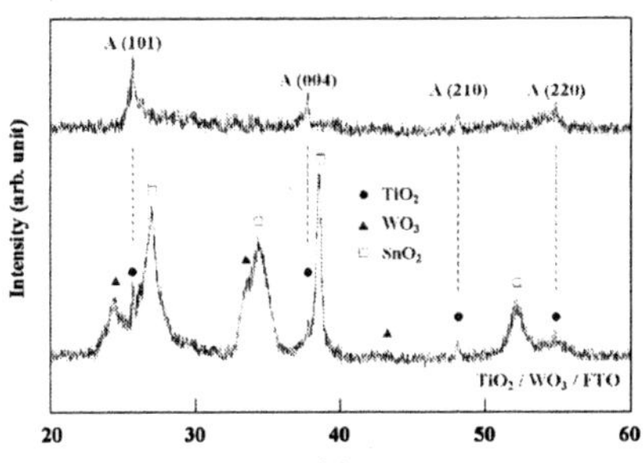

Figure 3. X-ray diffraction patterns of TiO_2 / WO_3 / FTO multi-layer structure.

Figure 4 shows transmission spectra of the single films and multi-layer structure. In the visible-light region, TiO_2 and SnO_2 films are transparent, however, the WO_3 film takes place absorption. The absorbance of the TiO_2 / WO_3 / FTO multi-layer structure in the visible-light region became lower compared with the WO_3 film, because the light scattering increases at the boundaries between the films and/or surface roughness due to multi-layer structure.

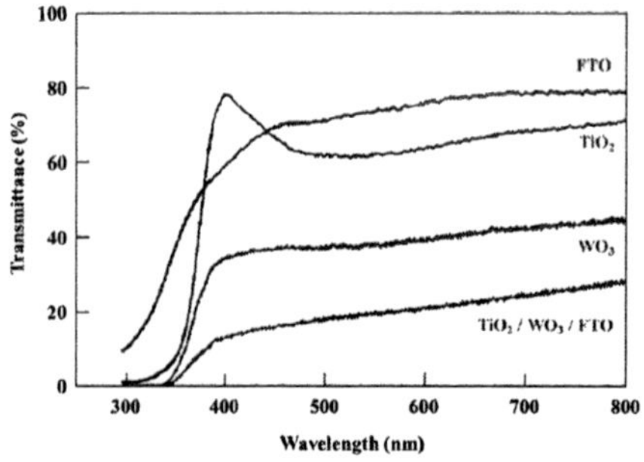

Figure 4. Transmission spectra of the single films and multi-layer structure.

Visible-light photocatalytic activity and spatial carrier separation

Figure 5 shows the visible-light photocatalytic activities for the various film structures. Only the TiO_2 film has no activity while the TiO_2 / WO_3 structure shows an activity. The activity of the structures with SnO_2 bottom layer becomes higher, and it is enhanced with increasing the fluorine concentration in the SnO_2 film. Because the visible-light absorption only takes place at the WO_3 layer, the visible-light photocatalysis of the multi-layer structures are due to carrier transfer from the WO_3 to TiO_2 films. Considering the position of band edges of the TiO_2 / WO_3 / SnO_2 structures as shown in Fig. 6, the electrons and holes generated in the WO_3 layer diffuse to the conduction band of SnO_2 and the valence band of TiO_2, respectively.

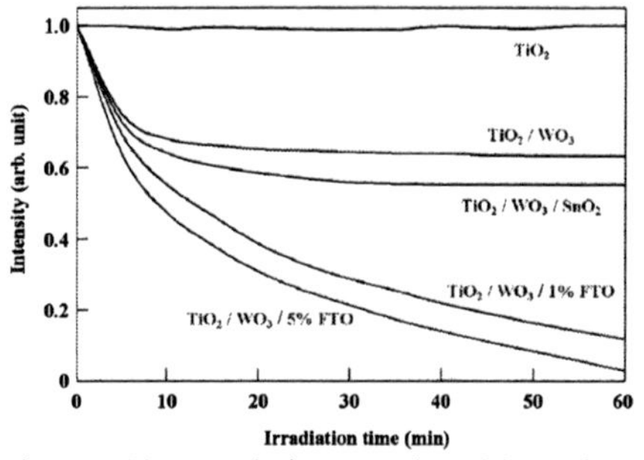

Figure 5. Photocatalytic properties of the various film structures.

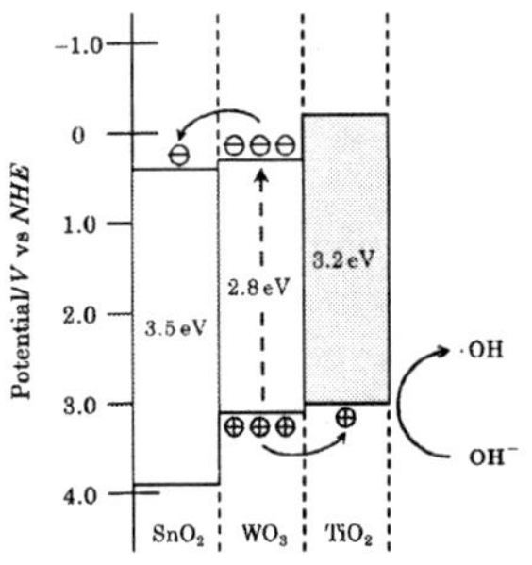

Figure 6. Position of energy bands of the three-level structure.

Therefore, the visible-light photocatalysis of the TiO_2 / WO_3 / FTO multi-layer structure is induced by the hydroxyl radicals caused by oxidation of adsorbed H_2O and/or OH^- by the holes existing in the valence band of TiO_2 layer. The role of the SnO_2 bottom layer scavenges the electrons diffusing from the WO_3 layer, and its effect is enhanced with increasing conductivity of the FTO bottom layer [4]. As a result, the FTO bottom layer contributes to the spatial carrier separation in the multi-layer structure.

These considerations were confirmed by measurement of the photocatalytic activities of SiO_2 inserted structures, that is the TiO_2 / SiO_2 / WO_3 and the TiO_2 / WO_3 / SiO_2 / 5% SnO_2 structures. As shown in Fig. 7, the carrier transfer from the WO_3 layer to the TiO_2 and FTO layers is inhibited. As a result, the photocatalytic activities of the TiO_2 / SiO_2 / WO_3 and the TiO_2 / WO_3 / SiO_2 / 5% SnO_2 structures become equally to those of the TiO_2 and the TiO_2 / WO_3 structures, respectively, as shown in Fig.8.

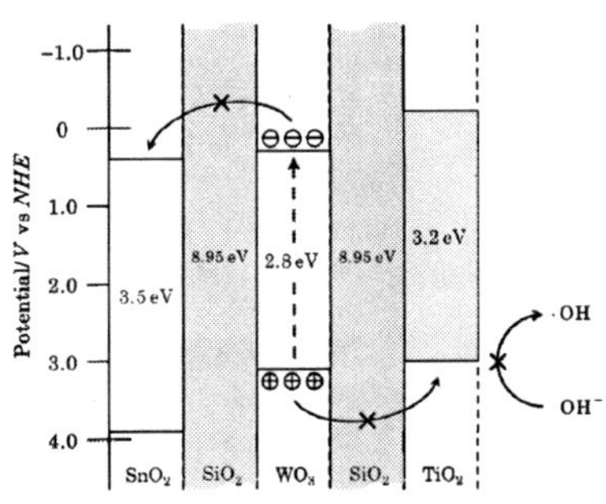

Figure 7. Position of energy bands of the SiO_2 layer inserted structures.

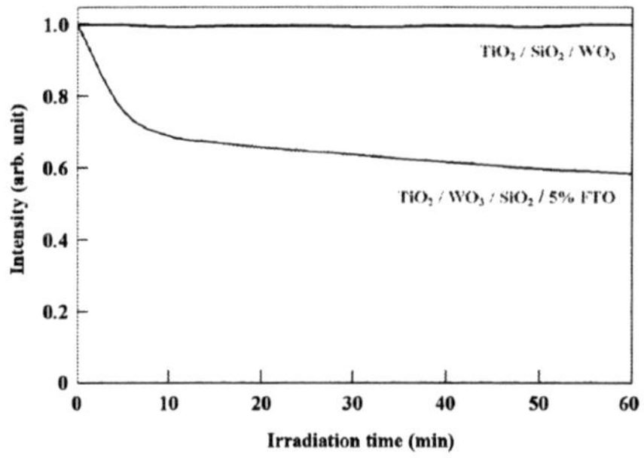

Figure 8. Photocatalytic properties of the TiO_2 / SiO_2 / WO_3 and the TiO_2 / WO_3 / SiO_2 / 5% SnO_2 structures.

CONCLUSIONS

The TiO_2 / WO_3 / FTO multi-layer structure with smooth surface was successfully prepared by SPD method. In this structure, the visible-light absorption takes place in the WO_3 layer and generated electrons and holes diffuse into the FTO and TiO_2 layers. The holes diffused into the valence band of the TiO_2 layer produce the hydroxyl radicals by oxidation of adsorbed H_2O and/or OH^-. The FTO bottom layer, which acts as a scavenger of electrons from the WO_3 layer, contributes to an effective spatial charge separation. As a result, the visible-light photocatalysis is enhanced in this structure.

ACKNOWLEDGMENTS

The authors wish to thank Mrs. Y. Ohmori and Y. Tsutsuura for their technical assistance.

REFERENCES

1. A. Fujishima, K. Hashimoto, and T. Watanabe, "TiO_2 Photocatalysis Fundaments and Applications," (BKC, 1999) Tokyo, Japan.
2. H. Shen, L. Mi, P. Xu,W. Shen, and P. N.Wang, App. Surf. Sci. **253**, 7024-7028 (2007).
3. H. Tada, A. Kokrubu, M. Iwasaki, S. Ito, Langmuir **20**, 4665-4670 (2004).
4. M. Oshima and K. Yoshino, J. Electronic. Mater. **39**, 819 (2010).

Mater. Res. Soc. Symp. Proc. Vol. 1492 © 2013 Materials Research Society
DOI: 10.1557/opl.2013.100

Photocatalytic TiO_2 Macroscopic Fiber Obtained through Integrative Chemistry

Natacha Kinadjian,[1,4] Mickael Le Bechec,[2] Thierry Pigot,[2] Fabien Dufour,[3] Olivier Durupthy,[3] Ahmed Bentaleb,[1] Eric Prouzet,[4] Sylvie Lacombe [2] and Rénal Backov[1]

[1] Université de Bordeaux, Centre de Recherche Paul Pascal, office 115, UPR 8641-CNRS, 115 Avenue Albert Schweitzer, 33600 Pessac, France
[2] IPREM- Institut des Sciences Analytiques et de Physicochimie pour l'Environnement et les Matériaux, Université de Pau et des Pays de l'Adour, Hélioparc-2 Av. du Président Angot, F-64053 Pau Cedex 09, France
[3] Chimie de la Matière Condensée-UMR 7574 CNRS-Université Pierre et Marie CURIE, Collège de France, Bâtiment C, 11 place Marcelin Berthelot, 75231 Paris Cedex 05, France
[4] Chemistry, ChemEng & Waterloo Institute of Nanotechnology, University of Waterloo, 200 University Avenue West, Waterloo (ON) N2L 3G1, Canada

ABSTRACT

Photocatalytic properties of titanium oxide depend on the material size and shape, which can favour a higher interaction between reactants and catalyst. Most of the studies reported until now, show that reducing size down to the nanoscale increases the photocatalytic efficiency. We demonstrate that a multiscale shape design, integrating surface roughness, particle shape, and material 1D processing and orientation, can favour photocatalytic properties in the solid-gas regime, especially mineralization (conversion into CO_2), when the material hierarchical 1D orientation is combined with unidirectional gas flow. Several materials with hierarchical structure were prepared and characterized. They have been tested for the photocatalytic mineralization of gaseous acetone, and compared with commercial catalysts. Our study reveals that a suitable combination of multiscale design can favour high mineralization.

INTRODUCTION

Titanium dioxide (TiO_2) is a widely used photocatalyst. Nanostructured TiO_2 exhibits superior photocatalytic activity compared to conventional bulk materials because of its high surface area over mass ratio [1]. TiO_2 photocatalytic ability was enhanced by preparing various morphologies of nanostructured TiO_2 [2,3]. Number of recent works especially reported the fabrication of one-dimensional (1D) TiO_2 photocatalyst [4]. The large surface-to-volume ratio, the enhanced adsorption of various reactants, and the favorable morphology for enhancing a longer reactant-catalyst contact, are expected to improve, overall, the properties of these materials [5]. Electrospinning offers several advantages to create fibers (1D structure) [6]. However, these fibers are randomly dispersed during the synthesis step, resulting in a direct loss of unidirectionality at the macroscopic scale. It is expected that enhanced photocatalytic performance could result from synergy between the 1D hierarchical catalyst structure, and optimized unidirectional gas flux in solid-gas photocatalytic experiments. Design of such a hierarchical anisotropy can be addressed through an Integrative Chemistry pathway, which allows for multiscale material processing [7]. Nanocomposite fibers can be produced as shown for the design and production of carbon nanotube [8], and vanadium oxide macroscopic composite fibers [9]. We studied in this report three types of titania nanoparticles (nanorods &

isotropic) that were synthesized, extruded as nanocomposite fibers, and characterized for their photocatalytic degradation and mineralization performance on acetone.

EXPERIMENT

All the products were purchased from Aldrich and used as received. The synthesis of TiO_2 nanorods was inspired from Sugimoto et al. [10]. The TiO_2 spherical particles were synthetized using the following process. Typically, HCl solution was first added to a 15%w TTAB solution. The pH of the solution was then increased up to 1.8 by addition of an ammonia solution. Titanium isopropoxide was then added under stirring. The resulting powder was washed with deionized water. Another synthetic path has been developed using Tergitol™ NP-10 in place of TTAB , all synthesis parameters remaining the same.

TiO_2/PVA fibers were prepared by co-extrusion. First, aqueous suspensions of TiO_2 were prepared at 2.5 wt% for titania nanorods,4 wt%, and 10.5 wt%, for spherical particles prepared with Tergitol and TTAB, respectively. Solid PVA (1.2g) was then added into the solution under stirring at 90°C for 1h. The fibers were generated by needle injection of the PVA/TiO_2 sols into a rotating containing a saturated salt solution of Na_2SO_4 at 45°C. The TiO_2/PVA fibers were meticulously taken out of the beaker by hand and allowed to dry in air. Finally, the fibers were washed once in water to remove the excess of salt. The as-synthesized fibers were then sintered at 450°C for 6 hours in order to remove the organic part while promoting crystallization.

Characterization

Scanning Electron Microscopy (SEM) observations were performed on a 6700F JEOL apparatus. Nitrogen adsorption method was performed with a Micromeritics ASAP 2010. Material roughness was evaluated from the determination of the Fractal surface Dimension (Ds), which can be deduced from the nitrogen isotherm adsorption curves [11]. Photocatalytic mineralization of gaseous acetone was carried out in a specially designed batch reactor based on a PTFE circulating loop where a photoreactor was inserted. The gas circuit consisted of a thermo-regulated pyrex reservoir tank connected to a Kel-F circulating pump. The gas was periodically sampled through a 10 $L.min^{-1}$ pump to a Varian 3800 GC equipped with a FID detector (for Volatile Organic Compound (VOC) monitoring) and a methanizer for CO_2 detection. Four fluorescent tubes (Philips TLD8W) with emission maximum at 366 nm were set upon the photoreactor to deliver an irradiance equal to 3.2 $mW.cm^{-2}$ UVA.

DISCUSSION

The TiO_2 sols can be extruded as composite PVA/TiO_2 macroscopic fibers (Figure 1). The PVA-TiO_2 dispersions were spun upon injection in a saturated aqueous solution of sodium sulphate salt (Na_2SO_4), to induce the coagulation of the PVA chains, via a salt effect [12]. This process leads to the formation of fibers in which the TiO2 particles are embedded within the coagulated PVA matrix. The extrusion process can be seen in Figure 1a and 1b. After performing extrusion, the fibers were extracted by hand (Figure 1c), washed, and air dried. These nanocomposite as-synthesized macroscopic fibers are made of TiO_2 nanoparticles, PVA, and some remaining sodium salt. These fibers were cut and placed in the reactor as their main axis will be parallel to the gas flow.

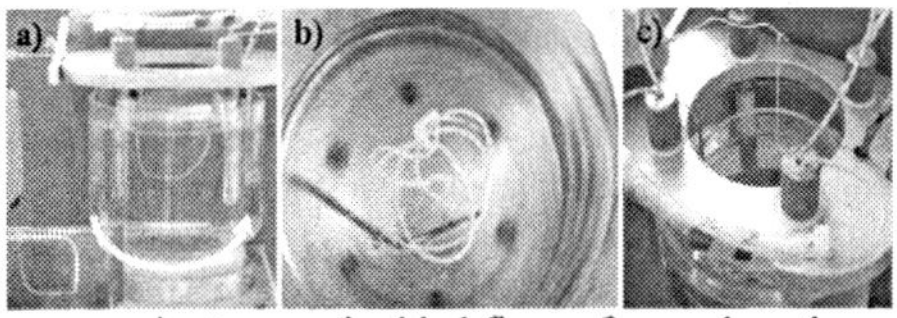

Figure 1. a) Extrusion process, bottom embedded figure focused on the needle out-flow, b) view from above once extrusion has been performed, c) fiber manual extraction from the beaker.

Overall, five different type of fibers were prepared and labeled hereafter X-Y-TiO_2 where X concern the % of particles weight fraction for the sol in use, and Y the nature of surfactant (2.5-TiO_2, 4-TTAB-TiO_2, 10.5-TTAB-TiO_2, 4-TERG-TiO_2 and 10.5-TERG-TiO_2).

The TiO_2 syntheses provide immediately the TiO_2 Anatase allotropic form and calcinations result only on creating inter-particles void space though PVA calcination. The fibers were treated at 450°C during 6 hours, in order to fully calcine the organic counterpart of the starting composite fibers, while promoting monophasic anatase crystallization. We observed that the diameter of the fibers varies with the initial concentration of the inorganic phase in the extruded sol. When the concentration in TiO_2 increases from 2.5 wt% to 4 wt%, then 10.5 wt%, the diameters increase as well from 230 μm, to 350 μm, and finally 500 μm, respectively. This evolution is the result of the variation in the sol viscosity, which balances the shear applied over the extruded jet.

It appears (Figure 2b) that the TiO_2 needles are randomly positioned within the fiber core, which precludes a similar shear-thinning effect. A close observation of the fiber surface (Figure 2b and 2d) reveals also the interparticle porosity, that ranks within the mesoporosity range (above 2 nm). The porosity was studied by nitrogen adsorption on calcined fibers (Figure 2e and 2f). All curves representing the istotropic particles are quite similar, we present the 4-TTAB-TiO2 as example figure 2f. They exhibit a rather straight adsorption curve, which can be described as a mix of type II-IV isotherms, along with a broad pore size distribution in the mesoporous range. Almost no hysteresis is observed. The pore size distribution, calculated with the BJH equation from the desorption curves, indicates that the pore size ranges between 6 nm and 15 nm.

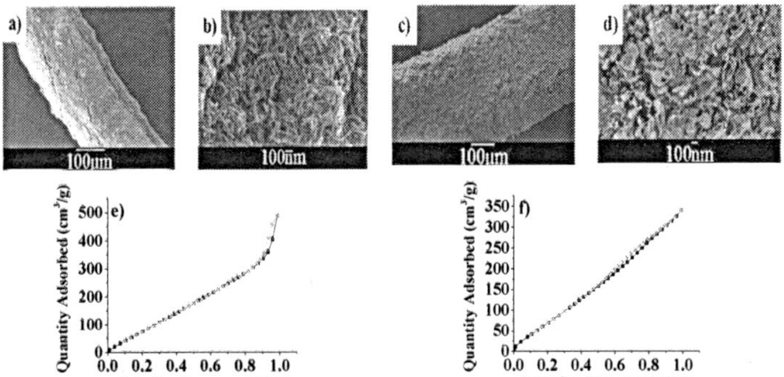

Figure 2. SEM images of 2.5-TiO_2 a) and b), 4-TTAB-TiO_2 c) and d). Nitrogen physi-sorption experiments. ■ Adsorption curves, ● desorption curves. e) 2.5-TiO_2, f) 4-TTAB-TiO_2.

The specific surfaces area of the titanium dioxide fibers are displayed in Table 2. All surface areas are relevant to mesoporosity. The specific areas are ranging from 100 to 700 $m^2.g^{-1}$. The lower surface areas are obtained with the fibers synthesized with the higher titanium oxide weight (10.5 wt%). We analyzed in parallel the surface roughness of these materials, as this feature can be an important factor in photocatalysis, for an optimized interaction between the pollutants and the catalyst. This analysis was carried out by using the nitrogen adsorption curve and calculating the surface fractal dimension Ds, taken as a descriptor of the surface roughness (Figure 3). We see in Table 1 that Ds becomes closer to 2, as the TiO_2 content increases. We assume that increasing the percentage of inorganic mater within the starting sol favors the stacking of the titania particles during the drying process, and increases the efficiency of sintering during the thermal treatment, which results in both lower porosity and surface roughness.

We probed the influence at different scales (1D micrometric fibers, isotropic or needle-like particles, surface roughness) of the hierarchical morphogenesis of TiO_2 materials onto their photocatalytic performances. Acetone was chosen as a test for Volatile Organic Compound (VOC) decomposition, because its photocatalytic mineralization is well documented [13,14]. In a first step, the photocatalytic device and the reaction conditions were settled with commercial TiO_2 fibers (Quartzel® PCO, Saint-Gobain Quartz). The performances of the TiO_2 fibers were then compared to those of this commercial material under identical conditions. Complementary experiments were made with the 90 µm powder from the grinded 10.5-TERG-TiO_2 fibers and the commercial powder "TiO_2 P25 EVONIK", used as a reference material.

Table 1. Specific surface area and surface roughness of the as-synthesized TiO_2 fibers obtained through Nitrogen physi-sorption experiments.

Fibers	BET surface $m^2.g^{-1}$	BJH surface $m^2.g^{-1}$	Total Pore Volume $mL.g^{-1}$	Ds (surface roughness)
2.5-TiO_2	520	450	0.76	2.37
4-TTAB-TiO_2	430	412	0.53	2.3
10.5-TTAB-TiO_2	106	120	0.16	2
4-TERG-TiO_2	700	652	0.82	2.31
10.5-TERG-TiO_2	190	214	0.26	2.095

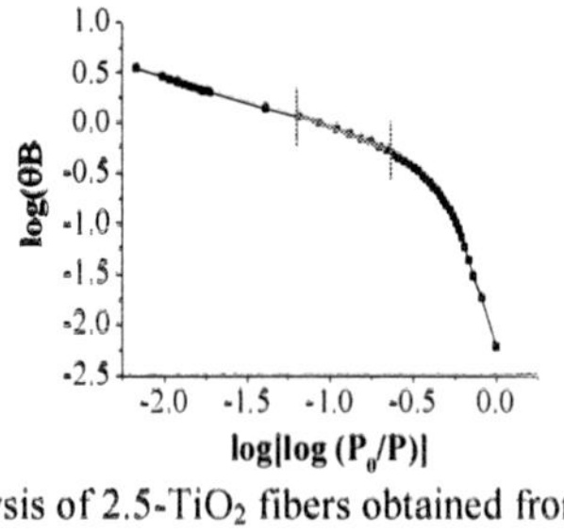

Figure 3. Surface fractal analysis of 2.5-TiO_2 fibers obtained from the adsorption curve of the isotherm .

These two types of powder were deposited as a slurry on the pyrex canister in the reactor. Mineralization is defined as the ratio of the number of moles of CO_2 evolved on the theoretical number of CO_2 (three times the number of degraded acetone). It should be emphasized here that the comparison between these media is not obvious since the composition of Quartzel® is unknown, their position in the reactors was not comparable. In Figure 4, the results are given in % acetone degradation/ g media relative to % mineralization after 400 minutes irradiation. The main result is that all the fibers present photocatalytic properties (Figure 4). The commercial materials have a higher degradation rate, which is not surprising since our fibers are too thick to allow a high light penetration.

Nevertheless, we observe that our fibers exhibit a higher mineralization rate (88-95 % degradation of acetone into CO_2) than those of the commercial products (82-84 %) (Figure 4). No by-product was detected within the sensitivity of the analytical device (0.1 ppmv), while high (80-95 %) but not complete mineralization was achieved. We assume that there is not any by-product formed which means that all the acetone was mineralized. We noted also in Figure 4 that the 10.5-TTAB-TiO_2 fibers, with the lowest surface roughness (Ds = 2) exhibited the lowest percentage of mineralization. The 2.5-TiO_2 fibers were found to be the less efficient in term of acetone degradation but the most efficient in term of mineralization percentage (100 %). No significant improvement for acetone degradation was observed with the 4-TTAB-TiO_2 fibers. With 10.5-TERG-TiO_2 and 4-TERG-TiO_2 the acetone degradation is of the same order of magnitude. These results confirm that the initial size of fibers (several hundreds of microns) do not allow for a complete access of light to titania, but the higher level of mineralization supports the concept of a higher residence time for the acetone by-products leading to a better mineralization.

In addition, the weaker efficiency of the 10.5-TTAB-TiO_2 material can be assigned to the lack of surface roughness, compared with other materials. This factor was checked by comparing the 10.5-TERG-TiO_2 with the same fibers crushed into 90 μm powder. The results obtained with the crushed fibers demonstrate that the degradation is improved by the surface effect due to a better irradiation, but the mineralization is reduced due to the disappearance of the shape factor that allowed for a higher residence time.

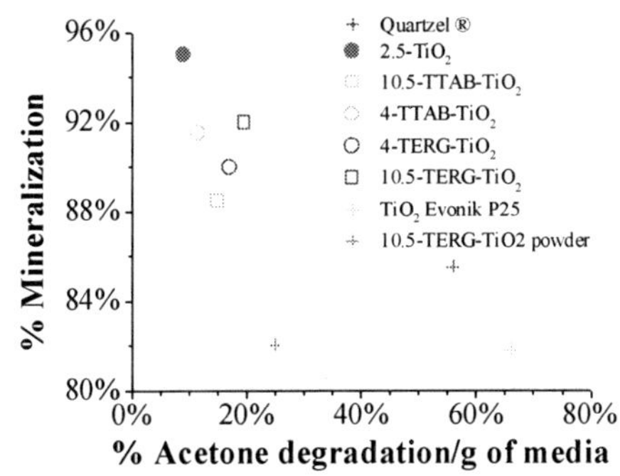

Figure 4. Comparison between acetone degradation (ppmv)/g media relative to % mineralization after 400 minutes irradiation for the five types of fibers, powdered 10.5-TERG-TiO_2 fibers, Quartzel®, and TiO_2 Evonik P25 powder under irradiation of gaseous acetone at 366 nm.

CONCLUSIONS

PVA/TiO_2 macroscopic fibers can be prepared through original integration between sol-gel process, extrusion and salt-induced flocculation. Upon calcination at 450°C in air, stable macroscopic fibers made of anatase nanoparticles, are generated. The thermal treatment allows for creating some mesoporosity by burning the original PVA matrix. These fibers can be used under co-axial flux photocatalysis experiment of gaseous acetone. The comparison of reaction kinetics with commercial media is difficult due to different mass, shape and photon absorption by the media. However, we can still see that higher mineralization rate is achieved with the fibers than with commercial Quartzel® fabric, especially with 10.5-TERG-TiO_2 fibers. As mentioned before, we used the photocatalytic tests to identify the influence of hierarchical structural parameters (fiber shape, particle shape, surface roughness) onto the photocatalytic efficiency of titania. It is obvious that the overall size of our materials cannot allow them to compete -at this stage- with optimized commercial catalysts. This study allowed us anyway to understand the influence of structural factors on taylor-made materials, which will be used in the future to optimize their photocatalytic efficiency. The main penalty which has to be considered for optimizing the as-synthesized fiber photocatalytic properties is the light access to TiO_2. It could be achieved by reducing the TiO_2 fibers diameter.

ACKNOWLEDGMENTS

The authors thank O. Babot for nitrogen physisorption data acquisition. NK thank European Union for financial support (Erasmus Mundus 2: IDS-FunMat consortium) and Waterloo Institute of Nanotechnology (WIN) for financial support.

REFERENCES

1. M. R. Hoffmann, S. T. Martin, W. Choi, D. W. Bahnemann, *Chem. Rev.* **95**, 69 (1995).
2. T. Kasuga, M. Hiramatsu, A. Hoson, T. Sekino, K. Niihara, *Adv. Mater.* **11**, 1307 (1999).
3. D. T. On, *Langmuir* **15**, 8561 (1999).
4. a) S. Madhugiri, B. Sun, P. G. Smirnitotis, J. P. Ferraris, *Microporous Mesoporous Mater.* **69,** 77 (2004) ; b) D. Li, Y. N. Xia, *Nano. Lett.* **3,** 555 (2003).
5. Y. Liu, D. D. Sun, J. O. Leckie, *Nano. Lett.* **7**, 1081 (2007).
6. R. A. Caruso, J. H. Schattka, A. Grenier, *Adv. Mater.* **13,** 1577 (2001).
7. a) R. Backov *Soft Matter* **2**, 452 (2006); b) E. Prouzet, Z. Khani, M. Bertrand, M.Tokumoto, V. Gyuot-Ferreol, J. F. Tranchant, *Micro. Meso. Mater.* **96**, 369 (2006).
8. B. Vigolo, A. Penicaud, C. Coulon, C. Sauder, et al., *Science* **290**, 1331 (2000).
9. L. Biette, F. Carn, M. Maugey, M.-F. Achard, et al., *Adv. Mater.* **17**, 2970 (2005).
10. T. Sugimoto, X. Zhou, A. Muramatsu, *J. Coll. Inter. Sci.* **259**, 53 (2003).
11. a) D. Avnir, M. Jaroniec, *Langmuir* **5**, 1431 (1985) ; b) E. Prouzet, C. Boissière, S. S. Kim, T. J. Pinnavaia, *Microporous and Mesoporous Materials* **119**, 9 (2009).
12. I. Sakurada in *Polyvinyl Alcohol Fibres* (Eds.: M. Dekker), International Fiber Science and Technology, New York, **vol. 6.**, (1985).
13. J. Peral, D. Ollis, *J. Catal* **136**, 554 (1992).
14. A. S. Besov A. V. Vorontsov, *Catal. Com.* **9**, 2598 (2008).

Mater. Res. Soc. Symp. Proc. Vol. 1492 © 2013 Materials Research Society
DOI: 10.1557/opl.2013.153

Novel protection solutions against environmental attack for light weight high temperature materials

Alexander Donchev, Michael Schütze

DECHEMA-Forschungsinstitut, Frankfurt am Main/Germany

ABSTRACT

The use of light weight structural materials such as titanium in transport systems like aero planes leads to a significant reduction in fuel consumption. However, titanium and its alloys cannot be used at elevated temperatures above 500°C for several reasons. Today aero engine compressors are made of a mixture of light Ti- and heavy Ni-alloys. The improvement of Ti-alloys to withstand the conditions in the high pressure compressor i.e. temperatures above 500°C would enable the manufacturing of a compressor from titanium as a whole with all its associated benefits. Intermetallic TiAl-alloys are another class of light weight materials for several high temperature applications. The use of TiAl as low pressure turbine (LPT) blades in the last sections of a large jet engine could save up to 150 kg of weight. In the last sections of the LPT the temperature is quite moderate (max. 650°C). The improvement of the high temperature capability of TiAl would allow its use in hotter sections of the engine with additional weight reduction. Similarly, the response performance of TiAl-turbocharger rotors in automotive engines would be much faster compared to the heavy Ni-based alloys used today. Furthermore higher rotation speeds are possible. Due to the novel so called fluorine effect the oxidation mechanism of TiAl can be altered. Fluorine-treated TiAl-components are protected by an alumina layer formed during high temperature exposure in oxidizing environments. This effect can be transferred to Ti-base materials if they are enriched with aluminum in a thin surface zone. The concepts and the results of high temperature exposure experiments of treated Ti- and TiAl-specimens are presented in this paper. They are discussed in the view of a use for real components.

INTRODUCTION

The increasing demands for reducing fuel consumption and hence the influence on the environment resulted in a search for more efficient automotive engines or jet turbines. The replacement of the relatively heavy-weight Ni- or Co-based superalloys currently applied by lighter materials would be an important step in this direction. Light weight intermetallic TiAl-alloys and optimized Ti-alloys could be used to reduce weight. Titanium is widely used as a structural material but not at elevated temperatures above 600°C [1]. Its use in hotter sections of e.g. a jet engine is not possible due to reduced oxidation resistance at high temperatures in oxidizing environments. Rutile is not protective anymore under these conditions in contrast to its passivation properties at low temperatures [2]. TiAl-alloys suffer from the same problem at temperatures above 750°C. They do not form a protective alumina scale during exposure in air like other aluminides (e.g. NiAl) due to similar thermodynamic stabilities of Ti- and Al-oxide [3]. A non-protective mixed $TiO_2/Al_2O_3/TiN$-scale is formed instead [4]. The use of TiAl as low pressure turbine blades in the last section of the engine could reduce the weight of each engine by more than 150 kg [1] furthermore the response behavior of TiAl-turbocharger rotors would be much faster [5].

It has been proven that the fluorine effect changes the oxidation behavior of TiAl-alloys at elevated temperatures. Small amounts of halogen at the metal/oxide interface lead to the formation of a slow-growing alumina barrier, which is protective against further oxidation even under complex conditions (humidity, thermal cycling) [6]. This effect works for TiAl-alloys with an Al-content higher than 40 at.% [7]. The enrichment of Al in a very narrow surface region of Ti-based materials leading to intermetallic TixAly-phases would allow the use of the fluorine effect for such alloys, too.

In this paper results from high temperature exposure experiments of untreated Ti- and TiAl-alloys are presented. They are compared with the results of Ti-specimens enriched with Al by the powder pack process and additionally F-treated and of TiAl-samples treated with fluorine exclusively. The results are discussed in view of the potential use of Ti- or TiAl-components at elevated temperatures.

EXPERIMENTAL

Specimens of commercially pure cp-Ti and a technical γ-TiAl-alloy (γ-MET; Ti-46.5Al-4(Cr, Nb, Ta, B)) were cut from sheets into coupons of 10 x 10 x 1 mm dimension. The coupons were ground with 1200 grit SiC-paper, cleaned with ethanol, rinsed with distilled water and dried in air prior to further use. The Al-enrichment was performed via the powder pack aluminizing process. For the F-treatment an F-containing polymer was sprayed in a homogeneous manner over all faces of the sample surface.

Isothermal tests were conducted in the temperature range from 600°C to 900°C in dry synthetic air for 120 h. The in-situ mass change was measured using a thermogravimetric balance (TGA). After the set temperature was reached the recording of the mass change due to oxidation was started. Hence, the mass loss of the F-treated samples due to the evaporation of the organic residues of the F-compound during the heating period was not considered in the TGA-results.

A second row of experiments were performed in a closed tube furnace isothermally under simulated Diesel atmosphere (N_2, < 10% O_2, < 10% CO_2, <1% SO_2) for 120h at 900°C. The furnace was heated up to 900°C under a steady stream of Argon. After this temperature was reached the samples were put into the hot furnace and the gas was switched to simulated Diesel atmosphere. The furnace was shut off after 120h and the samples were cooled down to room temperature (RT) under Argon.

Post experimental investigations included a first documentation of the surface with a Leica magnifier and a second analysis in the scanning electron microscope (SEM, Philips XL 40) with EDAX energy dispersive X-ray spectroscopy (EDX). Preparation of the metallographic cross-sections included galvanic Ni-plating. The cross sections were analyzed with a light microscope (LM) from Leica and SEM/EDX.

RESULTS AND DISCUSSION

The TGA-mass change curves of the untreated cp-Ti at 600°C and 700°C are presented in figure 1.The mass gain curves show a steady mass gain and the oxidation kinetics at 700°C is higher.

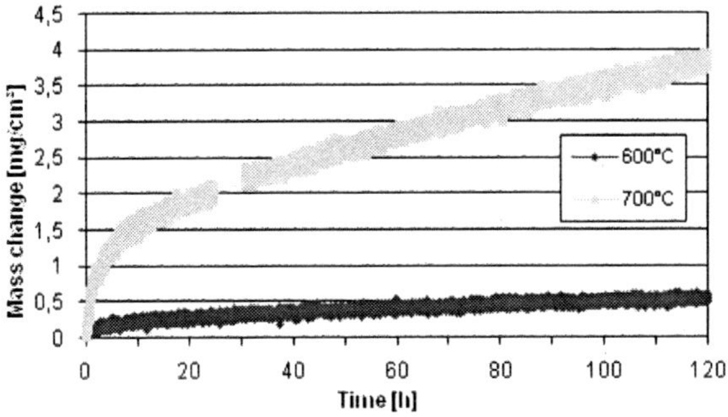

Figure 1: TGA-curves of the untreated Ti at 600°C and 700°C oxidized in air (The gap in the 700°-curve was caused by a computer failure)

Both samples exhibit a rutile layer which is less than 5 μm thick after 120h of oxidation at 600°C and roughly 10 μm thick after oxidation for the same exposure time at 700°C (fig. 2a, b). The oxide scales show cracks and pores and are therefore not protective. The oxygen and nitrogen can diffuse easily through the scales. EPMA-analysis revealed rutile formation plus inward diffusion of oxygen (fig. 3a) and nitrogen underneath the TiO_2-scale (fig. 3b). This is detrimental under thermocyclic and mechanical load because the oxygen or nitrogen enriched zone is more brittle than the substrate which can cause crack initiation and premature failure of Ti-components under service conditions.

Aluminization led to an intermetallic $TiAl_3$-layer directly after pack treatment [8]. This layer transfers into intermetallic Ti_yAl_x–phases due to interdiffusion with the substrate during HT-exposure and forms a thin protective alumina layer (fig. 4). This alumina layer hinders the inward diffusion of oxygen. The fluorine is needed to stabilize this layer [8]. Furthermore the intermetallic phases $TiAl_x$ ($x = 1$-3) have a very low oxygen solubility [9]. The elemental EPMA-analysis shows no oxygen within or underneath the intermetallic Ti_yAl_x-phases (fig. 5).

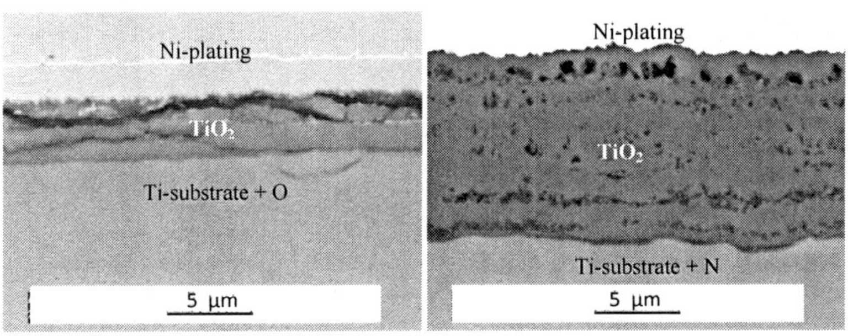

Figure 2a, b: SEM-images of the untreated Ti-samples after 120h of oxidation at 600°C (a) and 700°C (b)

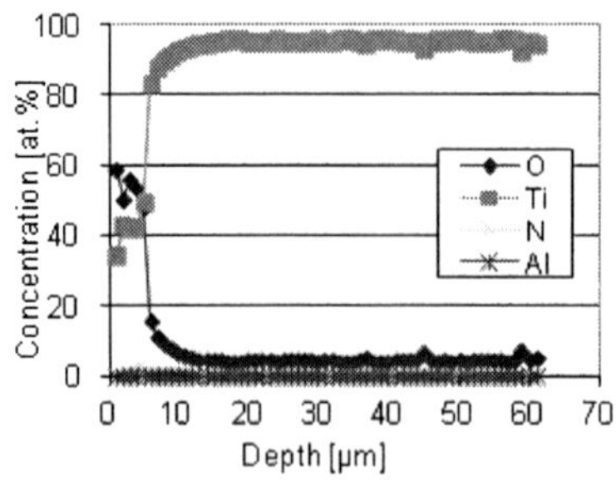

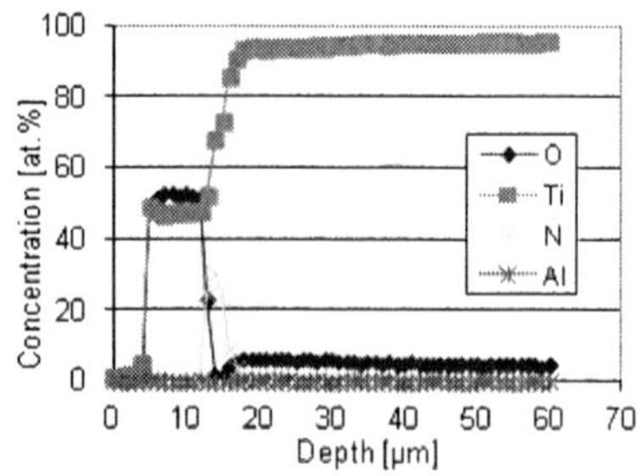

Figure 3 a, b: EPMA-elemental profiles of the untreated Ti-samples after 120h of oxidation at 600°C (a) and 700°C (b) in air

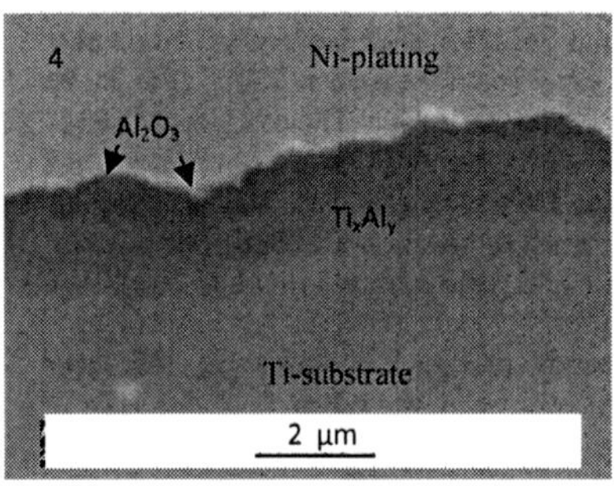

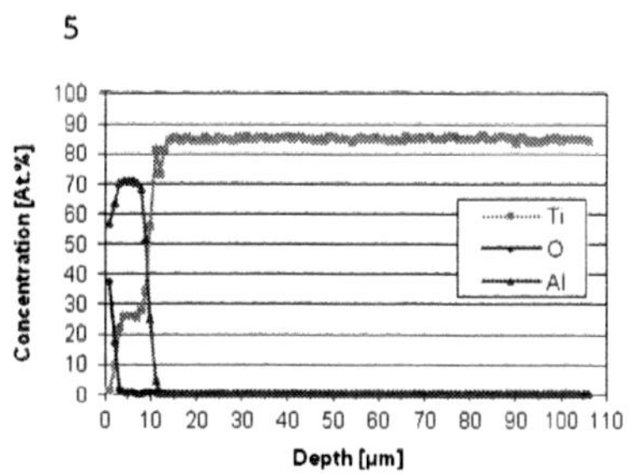

Figures 4 + 5: SEM-image (4) and EPMA-elemental profiles (5) of the combined Al + F treated Ti-specimen after oxidation at 700°C for 120h in synthetic air

TiAl is supposed to replace the heavy Ni-based alloys e.g. Inconel 713 as material for turbocharger rotors [10]. Therefore, exposure tests with the TiAl-alloy γ-MET were performed under simulated Diesel-atmosphere to investigate the behavior within an automotive engine. The surface of the untreated specimen is covered by TiO_2 crystals (fig. 6a), which were identified by EDX (fig. 6b). The Al-peak is caused by the Al_2O_3 within the mixed oxide scale because the energy was reaching some µm in depth (fig. 7a, b). This mixed scale is typical for untreated TiAl-alloys after high temperature oxidation.

In contrast, the F-treated TiAl-sample reveals a different surface structure. Only few crystals are visible on a rather smooth surface (fig. 8a). The EDX-spectrum exhibits a higher Al-peak indicating alumina formation which shows that the F-effect works also under these conditions (fig. 8b). Two very thin alumina layers are visible in the cross section (fig. 9a, b). An outer alumina barrier and an inner alumina enriched layer have been formed. The two layers are separated by the substrate. This explains the Ti-peak of the EDX-spectrum (fig. 8b).This double layered structure was found here for the first time. Usually only one thin outer alumina layer is found [6].The reasons for this phenomenon are under investigation and will be presented elsewhere.

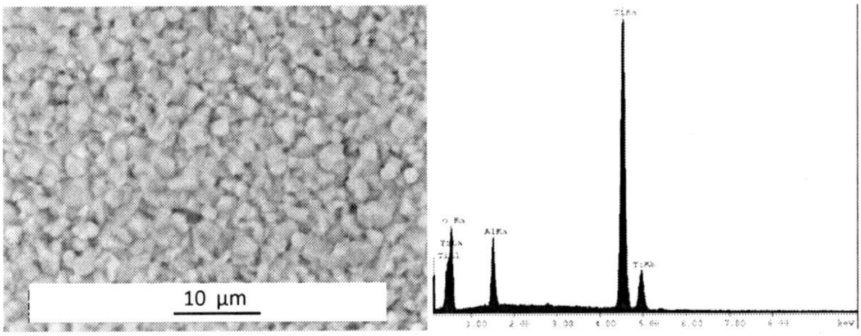

Figure 6a, b: SEM-image (a) and EDX-spectra (b) of the surface of the untreated γ-MET-sample was high after 120h at 900°C in simulated Diesel-atmosphere

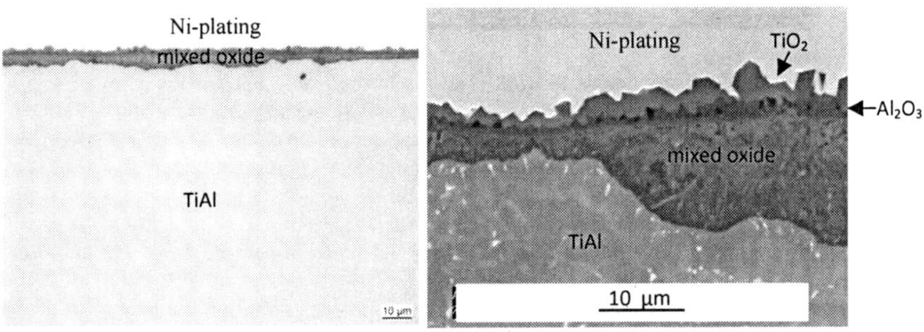

Figure 7a, b: LM- (a) and SEM-image (b) of the wavy mixed oxide scale of the untreated γ-MET-sample after 120h at 900°C in simulated Diesel atmosphere

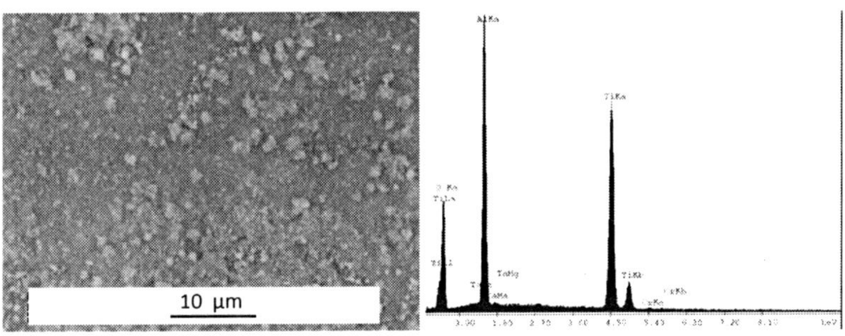

Figure 8a, b: SEM-image (a) and EDX-spectra (b) of the surface of the F-treated γ-MET-sample after 120h at 900°C in simulated Diesel atmosphere

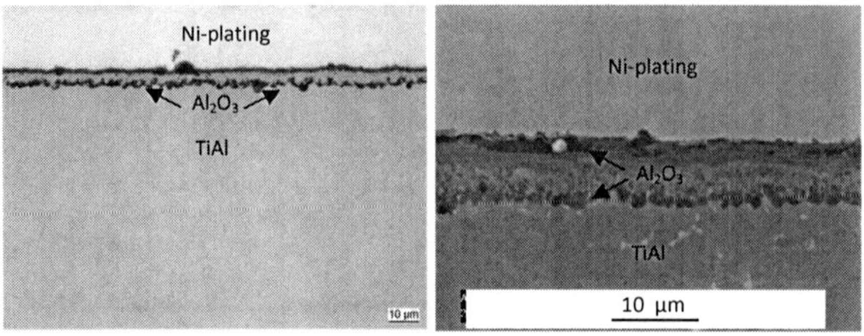

Figure 9a, b: LM- (a) and SEM-image (b) of the wavy mixed oxide scale of the F-treated γ-MET-sample after 120h at 900°C in simulated Diesel atmosphere

SUMMARY

The combined Al- plus F-treatment protects Ti by the formation of a protective alumina layer and suppresses the inward diffusion of oxygen and nitrogen which causes embrittlement. The intermetallic $TiAl_x$-phases have low oxygen solubility and serve as a reservoir for Al. Hence, Ti-components could be used in hotter sections of a jet engine. TiAl-alloys could be used for turbo-charger rotors after F-treatment because a protective outer alumina layer has formed instead of a non protective fast growing mixed scale. Both types of surface treatment would allow reducing the weight of the engines by changing from heavy nickel alloys to Ti-based light weight materials and therefore increase the efficiency.

REFERENCES

1. M. Peters, J. Kumpfert, C.H. Ward and C. Leyens, „γ-Titanium Aluminide Alloys: Alloy Design and properties“, *Titanium and Titanium Alloys,* ed. C. Leyens and M. Peters (WILEY-VCH, 2003) pp.333-350.
2. G. Lütjering and J.C. Williams, *Titanium* (Springer, 2003) pp. 48-50.
3. A. Rahmel and P. J. Spencer, *Oxid. Met.* **35,** 53-68 (1991).
4. C. Leyens, "Oxidation and Protection of Titanium Alloys and Titanium Aluminides", *Titanium and Titanium Alloys,* ed. C. Leyens and M. Peters (WILEY-VCH, 2003) pp.187-230.
5. M. Mayer and G. Krämer, *Turbochargers* (Süddeutscher Verlag, 2011) pp. 25-35.
6. A. Donchev, E. Richter, R. Yankov and M. Schütze, *Intermetallics* **14,** 1168-1174 (2008).
7. P. Masset and M. Schütze, *Adv. Eng. Mat* **10,** 666-674 (2008).
8. A. Donchev, A. Kolitsch, M. Schütze and R. Yankov, *Mat. Corros.* **62,** 695-698 (2011).
9. B,-J. Lee and N. Saunders, *Zeitschr. Metallk.* **88**, 152-161 (1997).
10. T. Tetsui, *Adv. Eng. Mat.* **3,** 307-310 (2001).

Mater. Res. Soc. Symp. Proc. Vol. 1492 © 2013 Materials Research Society
DOI: 10.1557/opl.2013.175

Synthesis of Novel Halogen-Free Phenol Based Polymers and their utilization as Flame Retardant in Polypropylene system

Ruchi Bakshi[1,2,4], Sethumadhavan Ravichandran[3], Weeradech Kiratitanavit[2], Jayant Kumar[1,4], Ramaswamy Nagarajan[2]
[1]Center for Advanced Materials University of Massachusetts, Lowell, MA 01854, USA
[2]Department of Plastics Engineering, University of Massachusetts, Lowell, MA 01854, USA
[3]Department of Chemistry, University of Massachusetts, Lowell, MA 01854, USA
[4]Department of Physics, University of Massachusetts, Lowell, MA 01854, USA

ABSTRACT

Increasing awareness of toxicity of halogenated flame retardants (FRs) has resulted in stringent regulations that require the elimination of their use in plastics. Consequently, there is considerable interest for the development of new varieties of high performance FRs. Here, we report the synthesis of new class of polyphenol based FR additives. Preliminary investigations indicate that these FR materials can possibly work through a combination of radical scavenging and char-formation. The thermal stability and char yield of polyphenol samples were investigated by thermogravimetric analysis (TGA). Thermal characterization indicated that these polymers produce significant amount of carbonaceous char upon combustion.
Keywords: Polyphenols, Flame Retardance, Carbonaceous char, Polypropylene

INTRODUCTION

Over the past few decades there has been an upsurge in the use of polymers with over 560 billion pounds of various types of polymers produced annually worldwide [1]. Their use is projected to rise at 5-7 % per year making them indispensable in a wide range of applications including clothing, furniture, electronic and optical devices, adhesives, coatings, foams, packaging materials, automotive body parts and biomedical applications. Most polymers are based on hydrocarbons and therefore are inherently flammable. Fire retarding/flame retarding (FR) additives are often added to polymer formulations in order to delay or reduce flammability and meet fire-safety regulations. FR can be classified very broadly into inorganic, organo-halogen, phosphorous based compounds [2]. Several commonly used organo-halogen flame retardants like Polybrominated diphenylether (PBDE) and metal hydroxides like antimony trioxide are toxic [3]. Some organo-halogen compounds are also environmentally persistent and have been shown to cause numerous human health problems [4]. These FR materials enter the environment in multifarious ways: during manufacture, product use, combustion, leaching and recycling [5]. Wide spread use of these highly toxic FR materials resulted in detailed evaluations and subsequent banning of some organohalogen FR additives in several states in the United States and in the European Union. This generates the need of less/non-toxic FR additives using safer starting materials and benign synthetic approaches without compromise in fire-retarding capability for existing and newer applications. The primary goal of the work presented here is to design and develop a novel class of non-halogenated FR based on polyphenols. These materials have the potential for exhibiting interesting

and unique combination of very good radical scavenging property and char forming capability, satisfying two fundamental requirements of a good FR material [6]. We envision that successful synthesis and incorporation of these materials in polymer blends will help realize their potential in FR applications.
In the present work, we report the synthesis and characterization of a series of polyphenol based FR materials. The phenolic monomers used for the synthesis of these polymers is presented in **Figure 1**. Fe(II)-N,N'- ethylenebis(salicylideneamine) (Fe-Salen) is used as the catalyst. The thermal properties (decomposition temperature and char yield) of these polyphenols have been evaluated using technique as Thermogravimetry Analysis (TGA).

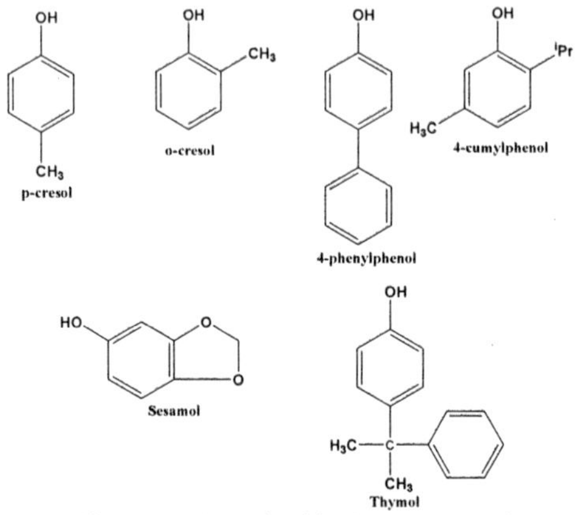

Figure 1. Phenolic FR's polymerized in the current study

EXPERIMENT

Materials

p-cresol (**1**), *o*-cresol (**2**), 4-phenylphenol (**3**), thymol (**4**), sesamol (**5**), 4-cumylphenol (**6**) were purchased from commercial sources such as Sigma Aldrich, Fluka, Alfa Aesar and Acros Organics respectively and used as received. Hydrogen peroxide was purchased from Alfa Aesar and Iron Salen was purchased from TCI America. Commercial polypropylene (PP, RJ470MO) with melt flow index (MFI- 2.8g/10 minutes at 190°C) was purchased from Borealis.

Typical procedure for polymerization of phenolic monomers using Fe-Salen [7]

Phenolic monomer (1 mmol) and Fe-Salen (0.1 mmol) were co-dissolved in tetrahydrofuran (5 ml) and the solution was stirred for 30 min. The polymerization was then initiated by incremental addition of a stoichiometric amount of hydrogen peroxide (3 % aq. solution, 1.8 ml, 1 mmol) over a period of 3 hours. After complete addition of H_2O_2, the reaction mixture was stirred for another 24 hours at room temperature. The solvent was evaporated under reduced pressure and the residue was washed thoroughly with 0.1 M HCl solution to remove Fe-salen. Isolated precipitate was then purified by Soxhlet extraction with methanol for 24 hours to remove un-reacted reactants. All samples were finally dried in a vacuum oven. In addition to the reaction performed at room temperature, reaction was also carried out at a lower temperature (4°C) for purpose of studying the impact of reaction temperature on the structure of synthesized polyphenols and consequently on their flame retardant properties [8].

Preparation of flame-retardant PP samples

Flame retardant masterbatches were prepared by mixing neat PP with 1, 5, 10 wt% of polyphenol based additive using a twin screw micro-compounding machine (DACA Instruments) at 195°C for 7-10 min with a mixing speed 100 rpm. Neat PP was translucent and it became opaque white and opaque brown with increasing polyphenols respectively. The composites were injected into standard testing bars for the tests of combustibility and a modified UL-94 Vertical burn tests was performed on the blends.

Characterization

Thermogravimetry analysis (TGA): Thermogravimetric analysis (TGA) was done using a TGA Q50 from TA instruments. The sample was heated from ambient temperature to 900°C at a rate of 10°C/min under constant nitrogen flow at the rate of 10mL/min.
Vertical Burning test: Vertical burning tests were conducted on 11 cm (= 4.33 inch) tensile bars (PP blended with phenol based polymer) with 1" inch long flame.

RESULTS AND DISCUSSION

Selection of monomers

In the quest to develop polyphenol-based flame retardants and to get an insight on the structural parameters that influence flame retardancy, we have selected a set of phenolic monomers having various substituents and polymerized them. The monomers are *p*-cresol (**1**), *o*-cresol (**2**), 4-phenylphenol (4-PhP) (**3**), 4-cumylphenol (**4**), sesamol (**5**), thymol (**6**) Earlier our studies have indicated that poly(*p*-ethylphenol) formed good char upon thermal decomposition. In order to study the effect of the type of substituent (alkyl, aryl or aralkyl) at the *para* position of phenol on the FR properties, we selected *p*-substituted phenols such as *p*-cresol (**1**), 4-phenylphenol (**3**) and 4-cumylphenol (**6**) as monomers for polymerization. *p*-cresol was selected as the rate of reaction of p-substituted phenols are quicker than other phenols. Additionally, electron donating methyl group activates the ring that contributes in the radical coupling process during polymerization. With the para position blocked the polymerization only proceed through the ortho position or through the phenolic oxygen. *o*-cresol was used to study the influence of substituent location on aromatic ring. 4-phenylphenol has phenyl ring at p-position, which should help stabilize the phenoxy radical. Upon heating the benzene rings can help in producing carbaneceous char. 4-cumylphenol, is an aralkyl phenol which is highly hindered and can produce stable radicals under fire. However, polymeriation of this monomer can be very challenging due to steric hindrance. Thymol, dialkyl-substituted phenol, is a natural mono-terpene phenol derivative of cymene and is found in oil of thyme, and is extracted from *Thymus vulgaris*. Thymol provides the basis for synthesis of dialkylated-phenol based polyphenol with a stable tertiary cumyl radical, which can facilitate radical formation during upon thermal decompostion. Sesamol is naturally occurring organic compound and is a component of sesame oil. Sesamol is a cyclic di-oxymethylene phenolic compound that provides an opportunity to synthesize a C-C coupled polyphenol.

Thermal decomposition behaviors

In order to understand the effect of synthesized polyphenol, the thermal degradation behavior and the amount of residual char obtained from TGA test carried out under nitrogen for various samples were compared. The thermal decomposition temperatures and char yields were tabulated in **Table 1**.

Table 1. TGA analysis of polyphenols synthesized using Fe-Salen at RT and LT

Polyphenol	T_d*(°C)				Char Yield (%) TGA	
	Reaction conducted at RT		Reaction conducted at LT		Reaction conducted at RT	Reaction conducted at LT
	T_{10}	T_{50}	T_{10}	T_{50}	T_{700} °C	T_{700} °C
Poly(4-cumylphenol)	277	483	266	474	21.0	39.0
Poly (4-phenylphenol)	344	766	326	725	57.0	65.0
Poly(o-cresol)	404	752	412	729	57.0	57.0
Poly(p-cresol)	350	786	317	766	61.0	58.0
Poly(sesamol)	247	356	248	586	8.0	23.0
Poly (thymol)	379	497	392	444	16.0	22.0

*T_d= Degradation Temperature, T_{10} = Temperature at 10% loss, T_{50} = Temperature at 50% loss

Key observations:

- In general, the polyphenols synthesized at low temperature exhibited better char forming ability compared to the polyphenols synthesized at room temperature. Low temperature polymerization has been reported to produce lower amounts of pummerer type ketones [8]. It is likely that low temperature favors formation of a greater promotion of C-C coupled products.
- In the case of p-substituted phenols such as 4-cumylphenol, 4-phenylphenol and p-cresol, the polyphenols synthesized at lower temperature (4°C) were found to be more thermally stable than those synthesized at room temperature.
- Among *p*-substituted phenols tested, poly (4-phenylphenol) and poly(p-cresol) exhibited better char yields compared to poly(4-cumylphenol), which indicates that substituent type of aryl and alkyl at *para* position of phenolic monomer is more favourable compared to combination of aryl and alkyl type (as in 4-cumylphenol) for obtaining polymers that produce higher char upon decomposition.
- Polyphenols derived from di-substituted phenolic monomers such as sesamol and thymol exhibited inferior char forming ability compared to mono substituted phenols.
- Poly(*o*-cresol) and poly(*p*-cresol) exhibited similar char forming abilities.

- TGA of all polyphenols synthesized at low temperature shows increased (15-20%) char yield when compared to the polymers synthesized at room temperature. This difference is because of formation of low molecular weight species (ketones) at room temperature. The oligomers sublime upon heating hence the amount of char produced is much lower.

Polyphenol based polymers as additives for polypropylene

Compounding: Polyphenol additives were compounded with PP using a twin screw micro-compounding machine (DACA Instruments). After optimization, the barrel was maintained at a constant temperature of 195°C for 7-10 min with a mixing speed 100 rpm for PP resins respectively. The compounded mixture was then compression molded and all further thermal characterization was performed on these molded specimens. Blends containing 1, 5 and 10 % of phenol based polymer were mixed and tested. Thermal decomposition, char yields of PP were studied by TGA and representative data from two of the samples is provided in **Table 2**.

Table 2. TGA analysis of polyphenols synthesized using Fe-Salen at RT blended with PP

Blends with PP	**T_d(°C)** (under N_2)	**Char Yield(%)** (under N_2)
Virgin PP	432	No Char
1% Poly(4-Phenylphenol)-PP	405	17.0
5% Poly(4-Phenylphenol)-PP	430	2.0
10%Poly(4-Phenylphenol) -PP	447	2.0
1% Poly(p-cresol)-PP	443	1.0
5%Poly(p-cresol)-PP	427	3.0
10% Poly(p-cresol)-PP	440	5.0

Key Observation:

- There is a significant increase in char yield in 5% and 10% loading for polyphenol FR blends.

Flammability Study using Modified UL-94 test

A modified UL-94 Vertical burn tests was performed on the blends and the results are tabulated in the **Table 3** given below

Materials	**Burning time per 4.33 inch of sample (second)**	**Results**
Neat PP	36	Oily drip, Faster flame propagation
PP + 1% Poly(4-PhP)	43	Oily drip
PP + 5% Poly(4-PhP)	55	Oily drip, slow flame propagation
PP + 10% Poly(4-PhP)	43	Oily drip, slower flame propagation than 1%

PP + 1% Poly(p-cresol)	37	Oily drip
PP + 5% Poly(p-cresol)	42	Oily drip
PP + 10% Poly(p-cresol)	50	Oily drip, slower flame propagation

CONCLUSION

A novel series of polyphenols with reasonable high thermal stability and good char forming capabilities have been synthesized and characterized. The effect of temperature on this biomimetic polymerization was studied. These results suggest that the reaction temperature plays a key role in the type and thermal stability of the products formed. Products with lower thermal stability are formed when the reaction is carried out at room temperature. The transition of the *ortho-para* dimer to Pummerer's type ketone structure occurs at room temperature and this is responsible for the formation of oligomers with poor thermal stability. However carrying out the reaction at lower reaction temperatures can minimize the formation of Pummerer's type ketone. Consequently, polyphenols synthesized at low temperature exhibited better char forming ability compared to the polyphenols synthesized at room temperature. These polyphenols were blended into polypropylene and the flammability of the blends (based on burn tests) was reduced by the addition of the polyphenols. More efforts are underway to understand the mechanism of the polymerization and the nature of the product that can provide superior thermal and FR properties.

ACKNOWLEDGMENT

We would like to acknowledge Dr. Ramanathan Nagarajan for the fruitful discussions and suggestions. We thank U.S Army Natick Soldier Research, Natick, MA for the financial support provided for this project 'GRANT NUMBER: W911QY-10-C-0174'.

REFERENCES

[1] Plastics News, January 7, 2008
[2] G.L. Nelson, Fire and Polymers V: *ACS Symposium Series*, 1(2007).
[3] K. D'Silva, A. Fernandes, M. Rose, Brominated Organic Micropollutants—Igniting the Flame Retardant Issue. *Critical Reviews in Environmental Science and Technology* **34**, 141(2004).
[4] C.A. de Wit, An overview of brominated flame retardants in the environment. *Chemosphere* **46**, 583(2002).
[5] S.L. Birnbaum, D.F. Staskal. Brominated flame retardants: cause for concern? *Environmental Health Perspectives*, **112**, 9(2004).
[6] S. Ravichandran, R. M. Bouldin, J. Kumar, R. Nagarajan, *Journal of Cleaner Production,* **19,** 454(2011).
[7] H. Tonami, H. Umaya, S. Kobayashi, *J.M.S.Ñ Pure Appl. Chem.* **A36**, 719(1999).
[8] X. Wu, W. Liu, R. Nagarajan, J. Kumar, L. A. Samuelson, A. L. Cholli, *Macromolecules* **37**, 2322(2004).

Efficiency and Fuel Switching

Mater. Res. Soc. Symp. Proc. Vol. 1492 © 2013 Materials Research Society
DOI: 10.1557/opl.2013.320

Mullite Formation in $Al_2O_3/SiO_2/SiC$ Composites for Processing Porous Radiant Burners

Daphiny Pottmaier, Jefferson J. Rosario, Marcio C. Fredel, Amir A.M. Oliveira, Orestes E. Alarcon

Mechanical Engineering Department, Federal University of Santa Catarina, Caixa Postal 476 Campus Trindade, 88040-900 Florianopolis, Brazil.

ABSTRACT

Use of porous ceramic burners for natural gas combustion is an optimum alternative to enhance energy efficiency and decrease emission of pollutant gases per generated power. Materials requirements for the operation of such porous burners are mainly thermal shock and chemical resistance and those can be reached with cellular ceramics. Mullite was theoretically identified among the best materials for this application; however, its potential was not properly explored yet. Even though mullite can be synthesized from different compounds and processing routes, control of final material characteristics is complicated mainly due to the formation of amorphous phase. In this work, using a technological approach mullite burners were processed by the replication method starting from different mixtures of $Al_2O_3/SiO_2/SiC$. Rheological study of the slurries has given additives content for the coating of the polyurethane sponges. After varying sintering temperatures up to 1600 °C and isotherm times for 12 h, microstructural aspects and product phases of the final composites were characterized in order to understand the influence of $Al_2O_3/SiO_2/SiC$ ratios in the formation of mullite phase and amorphous content.

INTRODUCTION

Porous medium combustion using radiant ceramic burners combined with natural gas is and optimum alternative to enhance energy efficiency of the combustion processes as it results in uniform and infrared heating with extension of lean flammability and higher burning rates with lower emission of pollutant gases [1]. That also means economic advantage as it lowers gas consumption per generated power. Combustion inside a porous structure requires specific properties from its constituent materials during operation. Materials for porous burners must resist principally to thermo mechanical stress due to high temperature gradients and corrosion effects depending on the fuel type, ratio and pressure. Additionally, low-priced raw materials are necessary for the fabrication of simple components in many domestic and industrial applications. Cellular ceramics suitably attend the required properties as a function of porosity percentage and morphology, porous distribution and type. Such characteristics may vary depending on the constituent materials (i.e. pure, mixture, composite) and their processing (e.g. replication, foaming, gelation). The most used materials in porous burner systems are: silicon carbide (SiC) due to good thermal transport up to 1350 °C and other outstanding properties [2], aluminum oxide (Al_2O_3) for temperatures up to 1900 °C [3], zircon oxide (ZrO_2) for temperatures up to

2300 °C [4], Fe-Cr-Al alloys used for lower temperatures and high thermal shock resistances [5]. They can also be fabricated with other materials such as corderite [6-8] and other approaches such as using aggregates or fibers [9].

Previous studies in our laboratory were performed using the replication method for processing of Li_2O-ZrO_2-SiO_2-Al_2O_3 (LZSA) glasses [10-12], ZrO_2 [13], Al_2O_3 [14], and recently for SiC-based foams [15-17] similarly to Zhu et al. [18-20]. From these analyzed porous systems, limitations assigned are thermal shock resistance of SiC and Al_2O_3, production costs of ZrO_2 and operation temperatures of LZSA. Requirements such as higher operation temperatures and lower pollutant emissions, for a more efficient application of the porous combustion technology, moved our efforts to study other materials. Taking into account properties, process, materials price and furnace set-up, mullite ($3Al_2O_3.2SiO_2$ to $2Al_2O_3.SiO_2$) burners can be considered an optimal alternative and still underestimated for such application [21, 22]. A detailed materials selection for the design of porous burners is given elsewhere [23]; considering advanced modeling of fluid flow, heat generation from chemical reactions, and heat transfer inside the burner structure. Formation of bulk mullite is a complex process that depends strongly on the synthesis method and its precursors. Both solid (sinter-mullite) and liquid (fuse-mullite) state reactions are influenced by powder characteristics, heat treatment, and Al_2O_3/SiO_2 ratio. The crystal structure of mullite is orthorhombic, space group Pbam, with a large range of solid solution, from $3Al_2O_3.2SiO_2$ to $2Al_2O_3.SiO_2$ is based on the substitution of $2Si^{4+} + O^{2-} = 2Al^{3+}$ on the tetrahedral sites [24]. Crystal morphology can be categorized in several types, such as granular, cuboidal, acicular, or needlelike; with aspect ratio varying according to the viscosity of the eutectic liquid and as a function of alkali content (Na_2O, K_2O) [25]. Addition of other compounds might be not effective in growing specific morphologies but can significantly increase its conversion rate. For example, only 20 % mullite was produced in the air heated sample, but 96 % mullite was reached in the presence of SiF_4 [26]. Furthermore, mullite can be formed by gaseous state processes such as chemical vapor deposition; for coating of SiC [27-29].

Finally, in this work, it was used an approach co-relating materials composition and microstructure of mullite-based systems to fabricate porous burners in order to resist high temperatures and harsh environments. Therefore, characterization results of the final composites herein are concisely discussed on the light of its materials technological application.

MATERIALS AND METHODS

Starting materials suppliers are listed in Table 1 together with composition of the $Al_2O_3/SiO_2/SiC$ suspensions. Commercial polyurethane (PU) sponges were impregnated with 7 ppi (Crest Foam Industries, United States of America). Particle size distribution of precursor powders was obtained with a ZS Zetasizer (Malvern Instruments, United Kingdom).

Table 1. Starting materials suppliers and composition of $Al_2O_3/SiO_2/SiC$ suspensions.

	Al_2O_3	SiO_2	SiC	colSil	BTE'	CMC''
Sample / Supplier	Alcoa		Treibacher Schleifmittel	Sigma Aldrich	Schumacher Insumos	EMFAL
(1) Al_2O_3 / SiO_2	72.0	28.0	-	65.3	2.0	0.75
(2) Al_2O_3 / SiC	72.0	-	28.0	-	1.5	0.5
(3) Al_2O_3 / SiO_2 / SiC	72.0	14.0	14.0	32.7	2.0	0.75
(4) SiC / Al_2O_3 / SiO_2	20.0	7.0	70.0	21.5	1.0	0.25

' – Bentonite '' – Carboxy Methyl Cellulose

Process flow of the replication method for the fabrication of the porous burners: (1) water suspensions of the precursor materials were prepared by ball milling using 1/3 of alumina balls, at 200 rpm for 24 hours; rheological studies of the suspensions performed with a Haake VT 550 (Thermo Fisher, United States of America). (2) Slurry immersion and impregnation was made using a home-made system. (3) Drying at ambient temperature for 24 hours. (4) Heating up to 850 °C at 1 °C/min for the extraction of the PU foam and organic additives in a muffle with exhaustion system (Irmaos Sanchis & Cia Ltda, Brazil).(5) Sintering at temperatures up to 1600 °C at 5 °C/min in a bottom loading furnace (Micropyretics Heaters International Inc., United States of America).

Thermogravimetric data were acquired with a Q 500 (TA instruments, United States of America) and dilatometry with a DIL 402C (Netzsch Group, Germany) in gas flow of 100 mL/min. Structural investigation was performed with XRD diffractometer X'Pert (Philips PANalytical, Netherlands) and microstructural analysis in a SEM XL30 (Philips, Netherlands).

RESULTS AND DISCUSSIONS

Starting Materials

Information related to the physical characteristics of the commercial powders is given in Table 2. High purity Al_2O_3 and SiC powders, confirmed by XRD patterns (not shown), without any additional treatment were used in the preparation of the composite materials. Particle size was also verified using an optical granulometry technique (not shown) in which Al_2O_3 presents a broad trimodal particle size distribution and SiC a sharper unimodal profile.

Table 2. Physical characteristics of precursor powders.

	Al_2O_3	SiC	SiO_2
Purity (wt.%)	99.6	98.5	-
Av. particle size (nm)	35	371	8
BET surface area ($m^2.g^{-1}$)	1.5	-	375.0

Figure 1. SEM micrographs of precursor Al_2O_3 (left) and SiC (right) powders.

Figure 1 shows a uniform morphology of the starting powders as resulting from SEM analysis. Al_2O_3 particles presented a facetted-barrel shape typical of its hexagonal corundum phase, while SiC particles are more angular with a plate-like shape and higher aspect ratio. Thus, anisotropy of SiC particles might play a role during materials packing and sintering in order to compensate its larger size and diffusion path compared to other precursors. In addition, thermal analysis of the commercial PU sponge (not shown) was performed and it revealed a profile of decomposition reactions with broad exothermic peaks between 250 °C and 350 °C and sharp peak at 525 °C; with gradual mass loss up to 900 °C and final carbonaceous residues of about 3.5 wt.%.

Slurries Behavior

Suitable rheological behavior of the $Al_2O_3/SiO_2/SiC$ suspensions (Table 1) for the impregnation of PU foams were obtained using Bentonite (BTE) as thickening agent and Carboxy Methyl Cellulose (CMC) as dispersant. The specification given by Studart et al. [30] was used as guideline, where viscosity should be 10 to 30 Pa.s at 5 s^{-1} shear rate and 1 to 6 Pa.s at 100 s^{-1} shear rate. Composition 4 was already studied using the same starting materials [15-17], thus it was promptly used in this work. The same systematic approach was followed to obtain the right additives proportion for the other compositions. As example, Figure 2 shows rheological profiles of the suspensions as a function of additive content for composition 1.

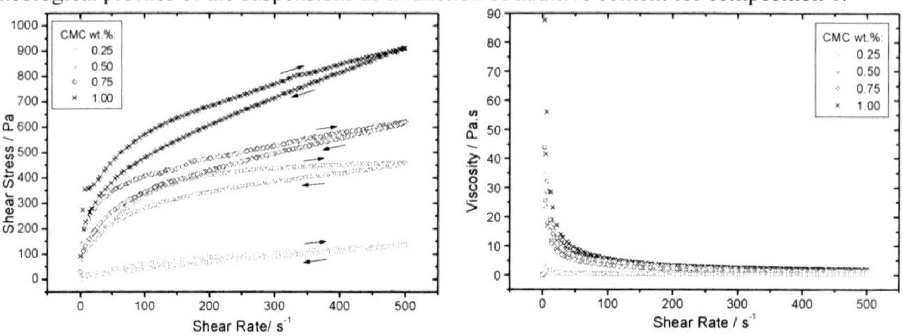

Figure 2. Al_2O_3-SiO_2 (1) slurry with BTE 2.0 wt.%: flow curves (a) and viscosity (b) per shear rate.

Figure 2a shows a pseudoplastic behavior of the slurry characterized by an increase in shear stress along with increase of shear rate. A thixotropic behavior can be characterized by the hysteresis on that curve. Figure 2b shows the behavior of viscosity with increasing shear rate, pseudoplastic and thixotropic behavior can also be observed in these curves due to the decrease of viscosity with increase of shear rate and the presence of hysteresis, as previously mentioned.

Sintering Profile and Structural Evolution

Figure 3 shows for comparison SEM micrographs of PU sponge as received, impregnated after burning and after sintering. It is observed a sharp skeleton of initial PU sponge with angular struts. After burning of the sacrificial template, the foam structure is formed by hollow struts together within holes resulting from the polymer evaporation. A more interconnected body is obtained for the sintered samples with rounded holes.

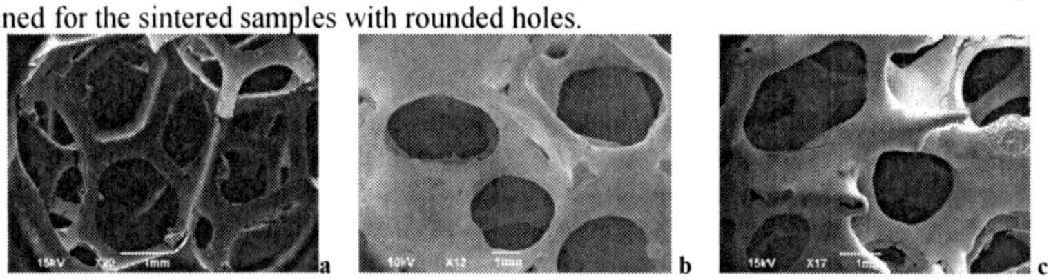

Figure 3. PU sponge as such (a), impregnated after burning (b) and after sintering (c).

Samples thermal behavior was studied coupling dimensional changes and weight changes (not shown) together with phase evolution (Figure 4) of the sintering process. Sintering profiles have shown a directly proportional relationship of colloidal silica content and both weight loss and dimensional contraction. Crystalline phases were identified together with amorphous content from XRD patterns of samples sintered at 1150 °C and 1600 °C.

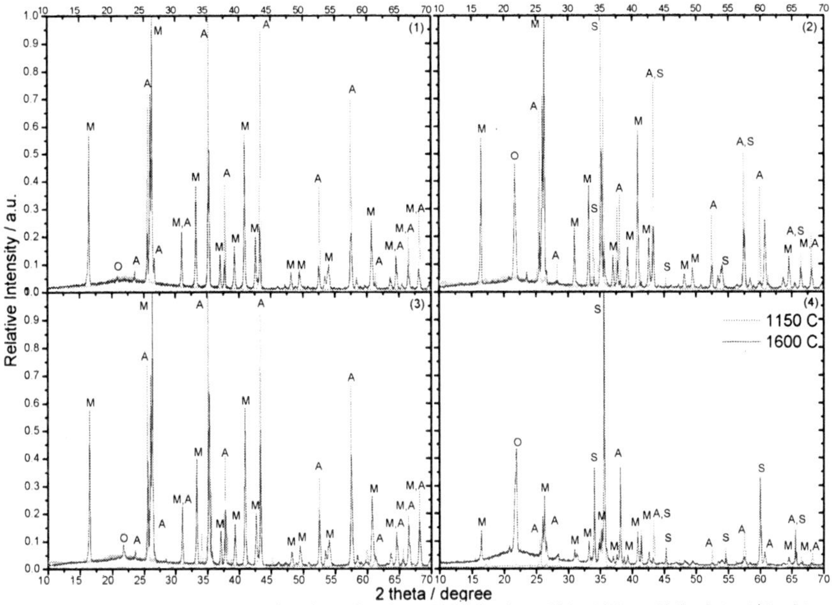

Figure 4. XRD patterns of: (1)Al_2O_3-SiO_2, (2) Al_2O_3-SiC, (3)Al_2O_3-SiO_2-SiC, (4)SiC-Al_2O_3-SiO_2. M= mullite, A= alumina, S= silicon carbide, O= silica.

Crystalline phases were assigned by their general names as solid solution may occur [31, 32], and therefore, dislocation of peaks positions are regularly observed for alumina and mullite phases. Alumina grains will disintegrate in the dissolution process due to Si^{+4} diffusion into Al_2O_3, starting in the glassy interface [33]. Thus, development of mullite and other aluminosilicates with similar crystal structure is accompanied by saturation of alumina phase. Such structural evolution is pronounced for samples 1, 2 and 3 with Al/Si ratio closer to mullite stoichiometry (Figure 4b). Another phenomenon occurring in parallel is the insertion of Al with both oxidation and amorphization of SiC [34]. This is suggested by the higher amorphous content of SiC-based samples 2 (Al_2O_3-SiO_2-SiC) and 4 (SiC-Al_2O_3-SiO_2). Additionally, the presence of small peaks suggests formation of other minor crystalline phases from the tripe junctions of Al_2O_3-SiO_2-SiC grains as recently reported in the literature [35].

Microstructural Aspects

Figure 5 shows SEM micrographs, in back-scattered electrons mode, of the polished surface for samples sintered at 1600 ºC under air atmosphere.

(a) (b) (c) (d)

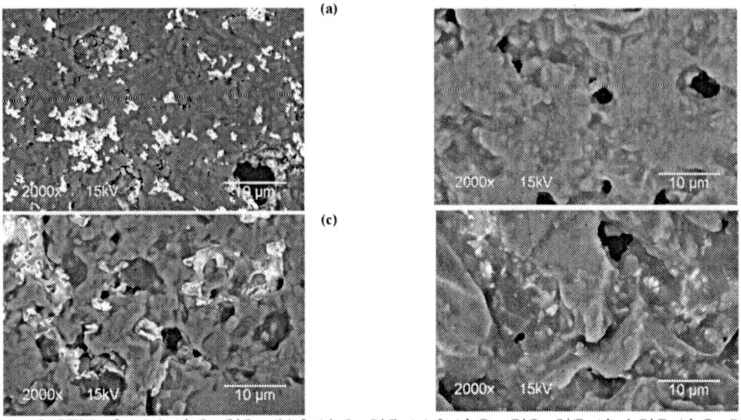

Figure 5. SEM of: (a) 1.Al_2O_3-SiO_2, (b) 2.Al_2O_3-SiC, (c) 3.Al_2O_3- SiO_2-SiC, (d) 4.SiC-Al_2O_3-SiO_2.

A more uniform microstructure consisting of homogeneously distributed phases is observed for composites 1 and 4. On the other hand, samples 2 and 3 presents an eye-visible gradient of phase from the surface to the center. Sharp facets of SiC particles are noticeably in samples 2 and 4, those with higher content of initial SiC (sample 4 > 2 > 3 > 1). Other feature observed by the SEM analysis is lower porosity and presence of intrinsic cracks for the composites with higher content of colloidal SiO_2 (sample 1 > 3 > 4 > 2).

CONCLUSIONS

Emphasis was given to find a simple and economical approach for processing mullite porous burners of high performances. Thus, formation of mullite and control of microstructural features were accompanied in order to obtain composites with the required properties.

Optimum rheology of the slurries for sponges impregnation was reached with additives content between 2.75 to 1.25 wt.% in a fixed 80/20 ratio of water to solid. Sintering profiles by dilatometry showed a contraction of 2.3 (sample 4) to 20.9 % (sample 1) and thermogravimetry a change in weight of + 2.0 (sample 4) to -18.0 wt.% (sample 1). Composite samples sintered at 1600 °C presented 26 (sample 4) to 82 (sample 2) wt.% of mullite phase and 35 (sample 2) to 54 (sample 1) wt.% of amorphous content. Accordingly, scanning electron micrographs of composites sintered at 1600°C have shown a phase gradient from shell to core as a function of colloidal silica (sample 1 > 3 > 4 > 2).

In addition, porous burners based in the Al_2O_3- SiO_2-SiC system are also being processed and characterized with the incorporation of residual ashes from coal combustion.

ACKNOWLEDGMENTS

Authors thank CAPES, CNPq, Petrobras for financial support; M. Petry, G. Senem, M. Leite, C. da Silva, R. Guimarares of CERMAT; J. F. Martins and M. Amorin of MAGMA; I. Mocellin and G. Hammes of LABMAT for technical assistance; and P. B. Prates of LCM for fruitful discussions.

REFERENCES

1. R. Viskanta, in *Handbook of Porous Media*, edited by K. Vafai (Taylor & Francis, New York, 2005).
2. A. Fuessel, H. Klemm, D. Boettge, F. Marschallek, J. Adler, A. Michaelis, IOP conf series in Ceram., Osaka (2011).
3. C. Tierney and A.T. Harris, Journal of the Australian Ceramic Society **45**, 20 (2009).
4. P.J. Elverum, J.L. Ellzey, D. Kovar, J. Mater. Sci. **40**, 155 (2005).
5. C.Y. Zhao, T.J. Lu, H.P. Hodson, J.D. Jackson, Mater. Sci. Eng. A **367**, 123 (2004).
6. A.A.M. Oliveira and M. Kaviany, Prog. Energ. Combust. Sci. **27**, 523 (2001).
7. F.A.C. Oliveira, S. Dias, M.F. Vaz, J.C. Fernandes, J. Eur. Ceram. Soc. **26**, 179 (2006).
8. S.M.H. Olhero, J.M.P.Q. Delgado, J.M.F. Ferreira, C. Pinho, Defect Diffus. Forum **273-276**, 814 (2008).
9. A.E.M. Paiva, P. Sepulveda and V.C. Pandolfelli, J. Mater. Sci. **34**, 2641 (1999).
10. J.J. do Rosário, R. Paiotti Marcondes Guimarães, M. Alves Leite, A.P. Novaes de Oliveira, M.C. Fredel, Mater. Sci. Forum **727-728**, 686 (2012).
11. M.W. Quintero, J.A. Escobar, A. Rey, A. Sarmiento, C.R. Rambo, A.P.N.d. Oliveira, D. Hotza, J. Mater. Process. Tech. **209**, 5313 (2009).
12. E. Sousa, C.B. Silveira, T. Fey, P. Greil, D. Hotza, A.P.N. de Oliveira, Adv. App. Ceram. **104**, 22 (2005).
13. S. Gómez, J. Escobar, O. Alvarez, C. Rambo, A. de Oliveira, D. Hotza, J. Mater. Sci. **44**, 3466 (2009).
14. V.M. Argüello, Master degree, Federal University of Santa Catarina, 2009.
15. G. Senem, Bachelor degree, Federal University of Santa Catarina, 2012.
16. C. Correa da Silva, Bachelor degree, Federal University of Santa Catarina, 2012.
17. M. Petry, Bachelor degree, Federal University of Santa Catarina, 2011.
18. X. Zhu, D. Jiang and S. Tan, Mater. Sci. Eng. A **323**, 232 (2002).
19. X. Zhu, D. Jiang and S. Tan, Mater. Res. Bull. **37**, 541 (2002).
20. X. Zhu, D. Jiang, S. Tan, Z. Zhang, J. Am. Ceram. Soc. **84**, 1654 (2001).
21. Ecocleantech, Standard Radiant Radiant Industrial Burners, (www.ecocleantech.nl).
22. F.A. Oliveira Costa, Mater. Sci. Forum **587-588**, 99 (2008).
23. J. Randrianalisoa, Y. Bréchet, D. Baillis, Adv. Eng. Mater. **11**, 1049 (2009).
24. D.X. Li and W.J. Thomson, J. Mater. Res. **6**, 819 (1991).
25. H.-Y. Lu, W.-L. Wang, W.-H. Tuan, M.-H. Lin, J. Am. Ceram. Soc. **87**, 1843 (2004).
26. A.J. Pyzik, C.S. Todd, C. Han, J. Eur. Ceram. Soc. **28**, 383 (2008).
27. O.R. Monteiro, Z. Wang, I.G. Brown, J. Mater. Res. **12**, 2401 (1997).
28. R.P. Mulpuri and V.K. Sarin, J. Mater. Res. **11**, 1315 (1996).
29. Y. Wang and W.J. Thomson, J. Mater. Res. **10**, 912 (1995).
30. A.R. Studart, U.T. Gonzenbach, E. Tervoort, L.J. Gauckler, J. Am. Ceram. Soc. **89**, 1771 (2006).
31. T. Kulkarni, H.Z. Wang, S.N. Basu, V.K. Sarin, J. Mater. Res. **24**, 470 (2009).
32. B. Saruhan, U. Voβ, H. Schneider, J. Mater. Sci. **29**, 3261 (1994).
33. Y.-F. Chen, M.-C. Wang, M.-H. Hon, J. Mater. Res. **19**, 806 (2004).
34. Z. Yang, H. Du, M. Libera, I.L. Singer, J. Mater. Res. **10**, 1441 (1995).
35. X.F. Zhang and L.C. De Jonghe, J. Mater. Res. **19**, 2510 (2004).

Mater. Res. Soc. Symp. Proc. Vol. 1492 © 2013 Materials Research Society
DOI: 10.1557/opl.2013.176

Highly efficient Synthesis of *per*-substituted amino-cyclodextrins under Microwave Irradiation in a closed Cavity

Giancarlo Cravotto*[1], Katia Martina[1], Marina Caporaso[1], Georgios Heropoulos[2] and László Jicsinszky[3]
[1] Dipartimento di Scienza e Tecnologia del Farmaco,
Università di Torino, Via P. Giuria 9, 10125 Torino, Italy.
[2] Institute of Organic and Pharmaceutical Chemistry
The National Hellenic Research Foundation, 11635 Athens, Greece.
[3] Cyclolab R&D Laboratory, Illatos út 7, H-1097 Budapest, Hungary.
*Author for correspondence: E-mail: giancarlo.cravotto@unito.it; FAX +39.011.6707687

ABSTRACT
Synthetic chemists are paying ever more attention to enabling technologies as a means to opening the paths towards the double goal of achieving high efficiency and meeting green criteria. Non-conventional techniques that create unique environments which promote selective modification have been investigated by many as part of the search for more highly efficient synthetic derivatization of cyclodextrins (CD). A few optimized microwave-assisted protocols have so far been developed for the preparation of selective *per*-alkylated aminocyclodextrin. In this work, a series of β-CD derivatives, whose primary hydroxyls were all replaced by amino groups, has efficiently been synthesized from *per*-(6-iodo-6-deoxy)-β-CD *via* nucleophilic substitution with amines under microwave irradiation in closed vessel (N_2 pressure). The reduction of *per*-(6-azido-6-deoxy)-β-CD to *per*-(6-amino-6-deoxy)-β-CD *via* catalytic hydrogenation has also been successfully carried out under dielectric heating.

INTRODUCTION
After more than 20 years of investigation, the use of dielectric heating to promote chemical reactions is now well established as a reliable technique which can be applied to a range of scales, from milliliters to kilograms [1]. The availability of professional microwave (MW) reactors (batch and flow), which provide control of all the main parameters, over a wide range of temperatures and pressures, often makes the critical conversions of even poorly reactive substrates feasible by cutting down on reaction time [2]. A comparison of processes performed under classic conductive heating and under MW irradiation is not a trivial task due to difficulties in understanding the true MW process temperature. The nature of the MW effect has been studied from the points of view of both reaction medium and mechanism. The effect should be stronger in apolar solvents and solvent-free reactions, with polar transition states and late transition states [3]. It is clear that the irradiation of MW absorbing materials (catalysts, carriers and reaction medium) can cause rapid volumetric heating and modify surface properties. This distinguishes the action of MW from traditional thermal treatments which are used in the preparation of catalysts and in reactions. The unique environments obtained under MW irradiation make this an useful enabling technique for the synthesis of tailor-made macromolecules, of which selective modified cyclodextrins (CDs) are a striking example [4]. CDs are natural cyclic oligosaccharides that are composed of seven α-(1-4)-linked glucose units and that feature a relatively rigid troncoconic structure. CDs have attracted worldwide interest from scientists in numerous host-guest recognition research fields [5]. Modified CDs have been the subject of intense study in recent years. The substitution of parent CDs can be an extremely

interesting process as the hydrophobic character of the interior cavity and its ability to form inclusion complexes remains unaffected, while solubility in aqueous media may be enhanced and binding tailored to specific needs. The synthesis of selective mono-, poly- and per-substituted CDs has often been limited by poor overall yield and time-consuming purification needs. In particular, full derivatization generally gives mixtures of polysubstituted products with a decreasing degree of substitution.

Per-(6-amino-6-deoxy)-β-CDs are an important class of derivatives because they are well suited to several applications (etc. biomimetic receptors) [6-9]. Amino-persubstitution on the primary face usually preserves the symmetry and the hydrophobic character of the CD cavity. *Per*-(6-amino-6-deoxy)-β-CD derivatives have attracted a great deal of attention because of their ability to act as ion channels in the mediation of anion over cation transport [6]. *Per*-(6-amino-6-deoxy)-β-CD has been employed as a starting material in the development of chemically and biologically stable macromolecule conjugates with the aim of studying and differentiating classical and non-classical ligand-activated estrogen receptor action [7]. Furthermore Stoddart, Zink *et al.* have described the development of pH-responsive nanoparticle-based drug delivery vehicles created by means of covering the surfaces of the nanoparticles with β-CD [8]. The reversible formation and hydrolysis of *per*-(6-amino-6-deoxy)-β-CD imine bonds that cover the surfaces of the nanoparticles was employed to release the large cargo molecules. Antitoxin activity has also been ascribed to complementary electrostatic interactions between cationic substituents at the primary hydroxyl group CD rim and the negatively charged amino acids at the inner surface of the protective antigen pore [9].

Modern applications of ultrasound (US) and MW irradiation promoted reactions are today not only widely used in classical organic synthesis, but have also made a major impact in macromolecule modification and in the field of polymer/material science [2b, 10]. In our recent papers, several CD functionalization processes, both classic and non-conventional MW / US procedures, have been compared. The results highlight the fact that these enabling technologies offer dramatic advantages in terms of yield and reaction time [4]. New selective chemical per-modifications of the primary face under non-conventional conditions have been investigated as a result of this background knowledge. One relevant task for this study was the simplification of the purification step which is even more problematic than normal in charged CD derivatives. The optimization of two prototypical synthetic schemes under non-conventional conditions are herein proposed. The efficient nucleophilic substitution of *per*-(6-iodo-6-deoxy)-β-CD with different amines and the catalytic hydrogenation of *per*-(6-azido-6-deoxy)-β-CD, described herein, stem from our previous MW-promoted protocols on CD derivatives.

EXPERIMENTAL

General Methods. Commercially available reagents and solvents were used without further purification. β-CD was kindly provided by Wacker Chemie. Reactions were monitored by TLC on Merck 60 F254 (0.25 mm) plates. Spot detection was carried out *via* staining with 5% H_2SO_4 in ethanol. NMR spectra were recorded with a Bruker 300 Avance (300 MHz and 75 MHz for ^{1}H and ^{13}C, respectively) at 25 °C. Chemical shifts were calibrated to the residual proton and carbon resonances of the solvent; DMSO-d6 (δH = 2.54, δC = 39.5), D_2O (δH = 4.79). Chemical shifts (δ) are given in ppm, and coupling constants (J) in Hz. ESI-mass spectra were recorded on a Waters Micromass ZQ equipped with an ESI source.

MW-promoted reactions were carried out in three professional reactors; the MicroSYNTH and the SynthWAVE by Milestone and the Discover by CEM.

per-(6-amino-6-deoxy)-β-CD (4)

per-(6-azido-6-deoxy)-β-CD (100 mg, 0.0763 mmol) was dissolved in 10 mL of MeOH and Pd/C (20 mg) suspended in H_2O (2 mL) was added. The reaction was carried out under magnetic stirring in a professional SynthWave MW (120 W) at 70 °C for 2.5 h with H_2 (10 bar). Solvents were partially evaporated and the solution was acidified with HCl to pH = 4.0. The solid was filtered, washed and the solution was lyophilized. The desired product was obtained in 91% yield (78 mg).

4 was an off-white powder; analytical data were in accordance with reported values [11].

General procedure for the synthesis of _per_-(6-alkylamino-6-deoxy)-β-cyclodextrin

per-(6-iodo-6-deoxy)-β-CD (0.0525 mmol) was dissolved in DMF (2 mL) and amine (5.25 mmol) was added. The reaction was carried out under magnetic stirring in a MW reactor (150 W) at 85 °C for 1 h. After concentration under vacuum to half volume and the addition of acetone (10 mL), a solid product was collected by filtration on a Hirsch funnel. The desired product was recovered without further purification. When the reaction was performed in the SynthWave MW reactor, 20 bar of nitrogen pressure were applied.

per-(6-benzylamino-6-deoxy)-β-cyclodextrin (5)

5 was a white powder; ^{1}H NMR (DMSO-d_6) δ= 7.27-7.10 (m, 35H, H-Ar), 5.76-5.73 (m, 14H, O(2)H, O(3)H), 4.83 (br s 7H, H-1), 3.60-3.50 (m, 28H, Ph-CH_2, H-5, H-3), 3.50-3.20 (m, overlapped with water, H-2, H-4), 2.61 (m, 14H, H-6) ppm; ^{13}C NMR (DMSO-d_6) δ 140.83 (C_{ipso}), 127.97 (C_{meta}), 127.65 (C_{ortho}), 126.41 (C_{para}), 102.14 (C1), 82.89 (C4), 72.91 (C3), 72.52 (C2), 70.62 (C5), 53.21 (Ph-CH_2), 48.34 (C6) ppm; MS (ESI): *m/z* calcd for $C_{91}H_{119}N_7O_{28}$ $[M + H]^+$ 1758.81 found 1758.59.

per-(6-allylamino-6-deoxy)-β-CD (6)

6 was a yellowish powder; ^{1}H NMR (DMSO-d_6) δ 6-20-5.85 (m, 21H, O(2)H, O(3)H,-C*H*=CH_2), 5.50-5.20 (m 14H, -CH=C*H*$_2$), 5.09 (s, 7H, H-1), 4.50-4.20 (m, 14H, $NHCH_2$), 3.80-3.00 (m, overlapped with water), 2.50-2.23 (m, overlapped with water) ppm; MS (ESI): *m/z* calcd for $C_{63}H_{105}N_7O_{28}$ $[M + 2H]^{2+}$ 704.85 found 705.33.

per-(6-butylamino-6-deoxy)-β-CD (7)

7 was a yellowish powder; ^{1}H NMR (DMSO-d_6) δ 6.10-5.86 (m, 21H, O(2)H, O(3)H), 5.07 (s, 7H, H-1), 4.20-3.80 (m, 7H, H-5), 3.80-3.30 (m, 21H, H-2, H-3, H-4), 3.30-2.80 (m, 28H, H-6, C*H*$_2$NH), 1.8-1.2 (m, 28 H), 1.02-0.73 (m, 21H) ppm; MS (ESI): *m/z* calcd for $C_{70}H_{133}N_7O_{28}$ $[M + 2H]^{2+}$ 760.96 found 761.41.

per-(6-methyl piperazino-6-deoxy)-β-CD (8)

8 was a white powder; ^{1}H NMR (DMSO-d_6) δ 5.96-5.85 (m, 14H, O(2)H, O(3)H), 4.90 (br s 7H, H-1), 3.70-3.20 (m, overlapped with water), 2.50-2.23 (m, overlapped with water) ppm; MS (ESI): *m/z* calcd for $C_{77}H_{140}N_{14}O_{28}$ $[M + H]^+$ 1710.00 found 1710.82.

DISCUSSION

The first synthetic method for the symmetric derivatization of α-, β- and γ-CD was described by Defaye and Gadelle [12a], where the selective replacement of all CD primary hydroxyl groups with iodine atoms was obtained using I_2 and Ph_3P in DMF. Based on this knowledge, we prepared *per*-(6-iodo-6-deoxy)-β-CDs (**2**), then successively treated them with NaN_3 in DMF to afford *per*-(6-azido-6-deoxy)-β-CD (**3**) in high yield (Scheme 1) [12b].

The preparation of *per*-(6-amino-6-deoxy)-β-CD derivatives has already been reported [11,13]. Synthesis *via* the *per*-(6-azido-6-deoxy)-β-CD derivative permits efficient Staudinger conversion to the *per*-(6-amino-6-deoxy)-β-CD. The latter is poorly soluble in all common solvents in the free base form (e.g. DMSO, DMF, H_2O), moreover, this route exacerbates the residual Ph_3PO drawback. As has already been described in the literature [14], the reduction of the epta-azido derivative by catalytic hydrogenation with Pd/C is a very difficult task. As a part of this work, a preliminary study investigated catalytic transfer hydrogenation, with hydrazine as the hydrogen donor, to avoid the presence of PPh_3O in the final product. Derivative **3** was heated under reflux in a solution of N_2H_4 in methanol/H_2O overnight and the desired amine **4** was collected in good yield as chloride salts. The presence of hydrazine in the final product, a serious drawback due to its high toxicity and instability, was detected by IR. This prompted the investigation of catalytic hydrogenation under MW irradiation in a professional MW reactor equipped with temperature and pressure control systems. A solution of *per*-(6-azido-6-deoxy)-β-CD in methanol/H_2O was irradiated at 70 °C in the presence of Pd/C under H_2 pressure (10 bar). The equipment guarantees homogenous and regular reaction mixture heating because the reaction vessel is immersed in an adsorbing medium, such as ethylene glycol. As depicted in Table I, the catalytic hydrogenation of **3** gave the pure final product in a 91% yield after 2.5 h of irradiation.

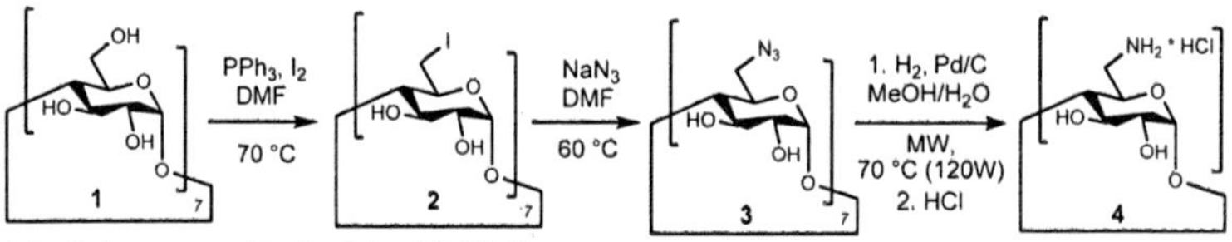

Scheme 1: Synthetic route to *per*-(6-amino-6-deoxy)-β-CD (**4**).

Table I Synthesis of *per*-(6-amino-6-deoxy)-β-CD

Entry	Reaction Conditions	Yield (%)	comments
1	PPh_3, NH_4OH, DMF, magnetic stirring, 25 °C, 18h	80 [13]	a
2	N_2H_4, Pd/C, MeOH/H_2O, reflux, 85 °C, 3h	85	b
3	Pd/C, H_2, MeOH/H_2O, SynthWave (120 W), 70 °C, 2.5h	91	-

a) The final product was isolated with PPh_3O impurities (7%), b) The product was isolated with N_2H_4 impurities as demonstrated by IR analysis.

The synthesis of β-CD derivatives *via* the direct amination of the *per*-(6-iodo-6-deoxy)-β-CD (**2**) was explored and provides easy access to a number of amino β-CD derivatives (see Table II). The aim of this study is to evaluate the influence of MW irradiation on nucleophilic substitution. Benzylamine was selected as the first model substrate for the optimization of reaction conditions. The heating of **2** at 75-80 °C in an amine excess, without solvent, for several days resulted in its quantitative conversion to the corresponding amino β-CD derivative [15]. When the reaction was performed in the presence of an amine excess, but in a minimum amount of DMSO or DMF solvent, it did not proceed to full conversion.

Per-(6-iodo-6-deoxy)-β-CD was reacted with excess benzylamine in DMF, all the reactions were performed under MW at 85°C. Interestingly, MW was more efficient than expected in cutting down reaction time and so the desired product was isolated by precipitation in acetone as the HBr salt after just 1 h of heating. Slight differences were observed in the monomode and multimode MW reactors. When the reaction was repeated with 50 eq of benzylamine, the reaction did not proceed to completion and a mixture of polysubstituted β-CD derivatives was isolated.

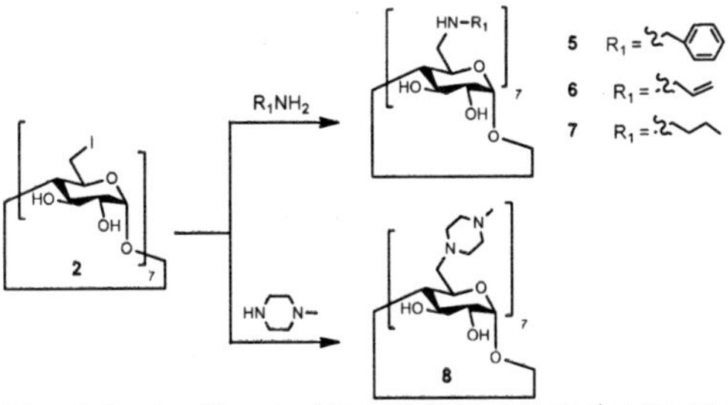

Scheme 2: Overview of the nucleophilic persubstitution of *per*-(6-iodo-6-deoxy)-β-CD (**2**) with different amines.

Table II: Nucleophilic persubstitution of *per*-(6-iodo-6-deoxy)-β-CD with different amines.

Entry	Amine	Amine (eq.)	Reaction time	Yield (%)
1	benzylamine	436	1h[a,b]	65[a], 68[b]
3	benzylamine	150	1h[b]	62
4	benzylamine	100	1h[b]	60
5	benzylamine	50	2h[b]	-[d]
6	1-methyl piperazine	436	1h[a,b]	48[a], 52[b]
7	1-methyl piperazine	100	1h[b]	55
8	1-methyl piperazine	436	1.5h[c]	55
9	allylamine	436	1.5h[c]	67
10	butylamine	436	1.5h[c]	63
11	aniline	436	2h[b]	-[d]

a) Multimode oven (MicroSINTH - Milestone) 85°C, 150 W; b) Monomode reactor (Discover - CEM) 85°C, 150 W; c) Multimode reactor (SynthWAVE - Milestone). N_2 (20 bar) 85°C, 150 W; d) The product was obtained as a mixture of polysubstituted β-CD derivatives

When repeated with aniline, the reaction afforded a mixture of polysubstituted β-CD derivatives. To assess the efficiency of the MW promoted procedure, the nucleophilic attack reactions of one secondary amine and of two volatile amines were performed in a professional MW reactor equipped with pressure control. As depicted in Table II, 1-methyl piperazine, allylamine and butylamine were chosen and successfully reacted in a SynthWave MW reactor (Milestone). The reactions were all performed under N_2 pressure (20 bar) which allowed for the use of a wider range of low-boiling reagents. This specific MW reactor, equipped with a rack with several test tubes, allowed the parallel synthesis of the 3 desired compounds to be carried out.

CONCLUSIONS

The irradiation of MW absorbing substrates and catalysts leads to unique environments that provide beneficial effects in terms of reaction rate and yield. Two efficient synthetic protocols that have directed the synthesis of monofacially functionalized CD bearing amino groups have been reported. The MW-assisted catalytic hydrogenation of *per*-(6-azido-6-deoxy)-β-CD afforded the pure desired product. The nucleophilic substitution of the *per*-(6-iodo-6-deoxy)-β-CD with different amines gave great advantages in yield and reaction time. The combination of facilitating technologies creates efficient and sustainable synthetic protocols.

ACKNOWLEDGMENTS

This work was funded by the Compagnia di San Paolo - University of Turin (Progetti di Ricerca di Ateneo 2011, grant D15E11001710003).

REFERENCES

1. a) Morschhäuser, R.; Krull, M.; Kayser, C.; Boberski, C.; Bierbaum, R.; Püschner, P.A.; Glasnov, T.N.; Kappe, C.O. *Green Process Synth.* **2012**, 1, 281-285. b) Roberts, B.A.; Strauss C.R. *Acc. Chem. Res.* **2005**, *38*, 653-661.
2. a) Mehtaa, V. P.; Van der Eycken, E. V. *Chem. Soc. Rev.*, **2011**, *40*, 4925-4936.b) Kappe, C. O.; Van der Eycken, E. *Chem. Soc. Rev.*, **2010**, *39*, 1280-1290.
3. de la Hoz, A.; Diaz-Ortiz, A.; Moreno, A. *Chem. Soc. Rev.*, **2005**, *34*, 164-178.
4. a) Martina, K.; Caporaso, M.; Tagliapietra, S.; Heropoulos, G.; Rosati, O.; Cravotto, G. *Carbohydr. Res.* **2011**, *346* (1 2677-2682; b) Boffa, L.; Calcio Gaudino, E.; Martina, K.; Jicsinszky, L.; Cravotto G. *New J. Chemistry* **2010**, *34*, 2013-2019 c) Cintas, P; Barge, A; Tagliapietra, S; Boffa, L; Cravotto, G. *Nature Protocols* **2010,** *5*, 607-616.
5. a) Szejtli, J.; Cyclodextrin and their inclusion complexes Academiai Kiado, Budapest, **1982**; b) Rekharsky, M. V.; Inoue, Y. *Chem. Rev.* **1998**, *98*, 1875-1917.
6. Madhavan, N.; Robert, E. C.; Gin M. S. *Angew. Chem. Int. Ed.* **2005**, *44*, 7584-7587.
7. Kim, H.-Y.; Sohn, J.; Wijewickrama, G. T.; Edirisinghe, P.; Gherezghiher, T.; Hemachandra, M.; Lu, P.-Y.; Chandrasena, R. E.; Molloy, M. E.; Tonetti, D. A.; Thatcher G. R. *J. Bioorg. Med. Chem.* **2010**, *18*, 809-821.
8. Zhao, Y.-L.; Li, Z.; Kabehie, S.; Botros, Y. Y.; Stoddart, J. F.; Zink, J. I. *J. Am. Chem. Soc.* **2010**, *132*, 13016-13025.
9. Diaz-Moscoso, A.; Mendez-Ardoy, A.; Ortega-Caballero, F.; Benito, J. M.; Ortiz Mellet, C.; Defaye, J.; Robinson, T. M.; Yohannes, A.; Karginov, V. A.; Garcia Fernandez J. M. *Chem Med Chem* **2011**, *6*, 181-192.
10. a) Xu, C.; Wang, J. , Wan, L.; Lin, J.; Wang, X. *J. Mater. Chem.* **2011**, *21*, 10463-10471; b) Franc, G.; Kakkar, A.; *Chem. Commun.* **2008**, 5267-5276; c) Puglisi, A.; Spencer, J.; Clarke, J.; Milton, J. *J. Incl. Phenom. Macrocyclic. Chem.* **2012**, *73*, 475-478.
11. Boger, J.; Corcoran, J.; Lehn, J.-M. *Helv. Chim. Acta* **1978**, *61*, 2190-2218.
12. a)Gadelle, A.; Defaye, J. *Angew. Chem. Int. Ed.* **1991**, *30*,78-80; b) Ashton, P. R.; Koniger, R.; Stoddart, J. F. *J. Org. Chem.* **1996**, *61*, 903-908.
13. a) Tagaki, W.; Yano, K.; Yamanaka, K.; Yamamoto, H.; Miyasaki, T. *Tetrahedron Lett.* **1990**, *27*, 3897-3900; b) Matsui, Y.; Tanemura, E.; Nonomura, T. *Bull. Chem. Soc. Jpn* **1993**, *66*, 2827.
14. Guillo, F.; Hamelin, B.; Jullien, L.; Canceill, J.; Lehn, J. M.; De Robertis, L.; Driguez, H. *Bull. Soc. Chim. Fr.* **1995**, *132*, 857-866.
15. Wang, Z.; Chang, L.; Klein, W. L.; Thatcher, G. R. J.; Venton, D. L. *J. Med. Chem.* **2004**, *47*, 3329-3333.

Mater. Res. Soc. Symp. Proc. Vol. 1492 © 2013 Materials Research Society
DOI: 10.1557/opl.2013.371

Enzyme-based Biohybrid Foams Designed for Biodiesel Production and Continuous Flow Heterogeneous Catalysis

Nicolas Brun,[1] Hervé Deleuze[2] and Rénal Backov[1]

[1] Centre de Recherche Paul Pascal, UPR 8641-CNRS, Université de Bordeaux, 115 Avenue Albert Schweitzer, 33600 Pessac, France.

[2] Université de Bordeaux, Institut des Sciences Moléculaires (ISM) UMR-5255-CNRS, 351 Cours de la Libération, 33405 Talence, France.

ABSTRACT

The one pot-synthesis and use of monolithic biohybrid foams in a continuous flow device reported inhere presents the advantages of covalent stabilization of the enzymes, together with a low steric hindrance between proteins and substrates, an optimized mass transport due to the interconnected macroporous network and a rather simplicity in regard of the column *in-situ* synthetic path. Those features, concerning transesterification (biodiesel production) enzyme-based catalyzed reaction, provide high enzymatic activity addressed with bio-hybrid catalysts bearing unprecedented endurance of continuous catalysis for a two months period of time.

INTRODUCTION

Modern societies will be confronted in the near future to an increasing requirement of fuels combined with a programmed diminution of easily extractable resources of fossil origin. In this context, biodiesels, which are fatty acid methyl or ethyl esters obtained by transesterification of vegetable oils has attracted considerable attention during the last decade[1]. Several processes for biodiesel fuel production have been developed, among which transesterification with alcohol using mineral acids[2], or bases[3] catalysts are the most common. Alkali-catalyzed alcoholysis, which is about 4000 times faster than its acid catalyzed counterpart, is commonly used in industrial processes[3]. Recently, enzymatic transesterification using lipases (triacylglycerol hydrolases), has become an interesting alternative for biodiesel fuel production, since the problems mentioned above can be circumvented[4]. In such context, it is necessary to develop processes insuring a complete recyclability and stability of enzymatic catalysts over the time, in order to reduce the impact of this cost increase down to competitive levels. Immobilization or entrapment of biocatalysts[5], onto or within porous materials using either physical adsorption,[6] covalent attachment[7], inclusion or encapsulation by sol-gel route,[8] represent an attractive and efficient approach to facilitate their use in continuous devices, and to compete with cheaper processes [9-13]. Therefore, the design of new functional porous materials to immobilize active biomacromolecules still present a challenge of both economic and ecological interests.

EXPERIMENTAL DETAILS

Syntheses. Tetraethylorthosilane (TEOS), tetradecyltrimethylammonium bromide (TTAB), (3-Glycidyloxypropyl)trimethoxysilane and dodecane were purchased from Fluka (St. Louis, MO). Lipase from Thermomyces Lanuginosus (solution, ≥100,000 U/g), carthame oil (Purified Safflower Oil), oleic acid, glyceryl trilinoleate (98 %), ethyl linoleate (≥99 %), linoleic acid (≥99 %), n-heptane, ethanol and 1-butanol were purchased from Sigma Chemical (St. Louis, MO). Other chemicals and solvents used in this study were of analytical grade or HPLC grade. Hybrid monolithic column synthesis: Typically, 5 g of TEOS was added to 16 g of a TTAB aqueous solution at 35 wt. % previously acidified (7 g of HCl), and hydrolysis was allowed to perform until a monophasic hydrophilic medium is obtained. The oily phase constituted of 35 g of dodecane was then emulsified drop by drop into the hydrophilic continuous phase using a mortar. The emulsion was then allowed to condense for 1 week at room temperature in a PTFE pipe (L = 251.0 mm ; r = 9.65 mm) confined in a stainless steel HPLC-type column lined with a PTFE tube (7 mm ID diameter, 300 mm useful length). The as-synthesized monolith was washed four days in a tetrahydrofuran/acetone mixture (at 50 vol %) under continuous flow (0.1 mL.min-1) to extract the oily (dodecane) phase from the monolith. Then, epoxy groups were grafted by a closed-loop continuous flow impregnation (0.1 mL min-1) of a 0.05 M solution of (3-Glycidyloxypropyl)trimethoxysilane (Glymo) in chloroform (200 mL) for three days. The hybrid silica-based column is then sucessesively washed with chloroform, acetone and water (0.5 mL min^{-1}). Transesterification catalysis reactions using Thermomyces lanuginosus crude lipase-based monolithic column: Crude lipase from Thermomyces lanuginosus (4 g; ≥ 100,000 U/g) was dispersed into distilled water (200 mL). The mixture was first stirred for an hour at room temperature. The hybrid column was impregnated by this solution in a closed-loop continuous flow process (flow rate: 0.1 mL min-1) during five days at room temperature. The as-synthesized bio-hybrid column is let two weeks at 4°C in the static enzymatic aqueous solution. After, the column was continuously washed with distilled water until disappearance of absorbance by Bradford measurement. In a second step, the bio-hybrid column was activated with a 5% w./v. glutaraldehyde aqueous solution (200 mL) in a closed-loop continuous flow process (flow rate:0.1 mL.min-1) during three days at room temperature. The activated biohybrid column was then functionalized with a new enzymatic aqueous solution (4 g TL crude lipase in 200 mL distilled water) by continuous flow impregnation (flow rate:0.1 mL.min-1) during five days at room temperature. The column was then continuously washed with distilled water, until disappearance of absorbance by Bradford measurement, and with heptane (flow rate: 0.1 mL min-1). The transesterification of crude safflower oil from Carthamus tinctorius seed (38 wt. %) by ethanol (12 wt. %) was performed in heptane (50 wt. %; 40 °C) at a 0.1 mL min-1 flow rate. Conversion of glyceryl trilinoleate was monitored using HPLC.

Characterizations. Scanning electron microscopy (SEM) observations were performed with a Hitachi TM-1000 apparatus operating at 15 kV. The specimens were sputtered with Au/Pd prior examination. HPLC analytical system: The analytical system consisted of a 600 solvent delivery system, manual injector (Waters, Milford, MA, USA). The compounds were separated on an Atlantis DC18 (4.6mm×150mm, 5μm) column with an Atlantis DC18 guard column (Waters). The column was operated at room temperature. Empower software (Waters) was used for data acquisition and processing. Standards were dissolved in tert-butyl methyl ether (MTBE). All solutions were filtered through a 0.45 μm membrane and degassed before use. The flow-rate was 1 mL min^{-1} and 20 μL of samples were injected. For the esterification catalyzed reactions a refractometer 410 (Waters, Milford, MA, USA) was used for detection. The mobile phase was acetonitrile (grade HPLC), isocratic elution. For the transesterification catalyzed reactions the detector in use was an ultraviolet diode array 996 (Waters, Milford, MA, USA). The maximum absorbance was at $\lambda = 204$ nm. A gradient elution was used. Solvent A: acetonitrile, solvent B: tert butyl methyl ether. A: 4 min 100 (v/v) isocratic, A-B: 2 min 70/30 (v/v) gradient, A-B: 10 min 70/30 (v/v), A: 0.5 min 100 (v/v) gradient. Then, the column was equilibrated under conditions mentioned above for 10min.

DISCUSSION

Continuous flow monolithic microreactors offer the possibility to circumvent some drawbacks associated with the use of batch reactors ; such limited productivity (low TON), complex control of heat transfers, high maintenance cost, security concerns, etc. In the present work, silica-based biohybrid monolithic foams have been synthesized in situ in a PTFE-coated stainless steel chromatography column by combining the sol-gel process with the concentrated emulsion soft templating approach (Figure 1a).

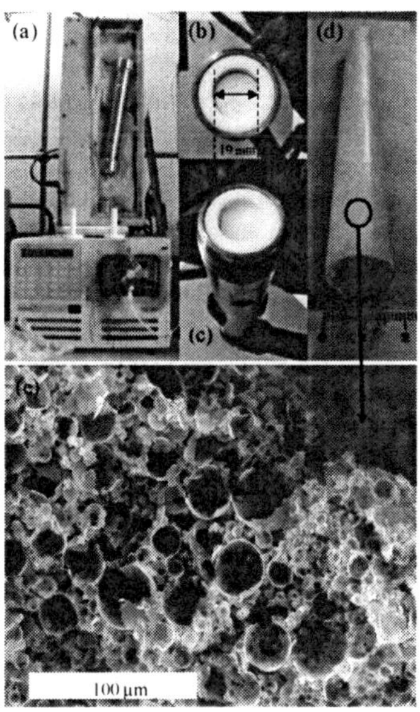

Figure 1. Pictures of: (a) the continuous flow set-up; (b, c) the as-synthesized column after 60 days of continuous flow transesterification catalysis within the stainless steel canister; (d) the biohybrid macrocellular column extracted from its canister after 60 days of continuous flow esterification catalysis. The column used in this study is labelled Col[C-TL-lipase]@gGlymo-Si(HIPE).

In order to covalently anchored the enzymes onto the inner monolith walls, epoxy groups have been grafted using a continuous flow impregnation with (3-glycidyloxypropyl)trimethoxysilane as organosilane precursor. Then, commercial crude lipases have been immobilized by continuous flow impregnation in distilled water (at pH = 6.5). Epoxy groups are very attractive candidates to generate short and soft covalent spacer arms and to react with different nucleophiles presents on the surface of enzyme molecules, as for instance amino, thiol, or hydroxyl groups [13, 14]. This feature allows reaching a good compromise between stability of the grafting and mobility of the protein, allowing, as observed in several cases, an exaltation of the immobilized enzyme activity [14]. Finally, after exhaustive continuous flow washings, the column containing the monolith, called Col[C-CR-lipase]@gGlymo-Si(HIPE), was disposed in the oven, and connected to a pump to be used in continuous flow catalysis, (See supporting information for the column characterization details over the micro-meso and macroscopic length scales). As can be seen in Figure 1b-c, after 60 days of catalytic use, the Col[C-CR-lipase]@gGlymo-Si(HIPE) monolith presents a low shrinkage (around 8 %) demonstrating the absence of a significant preferential flow circulation between the inner surface of the column and the outer surface of the monolith during the use (Figure 1d). Moreover, considering the low flow rate used in this study, (0.1 mL min-1), together with the bottom-up flow direction restricting gravity effect, and the relatively high viscosity of the moving fluid, creeping motion (Stokes flow) can be considered. This situation, characteristic of microfluidic systems, is favorable to a good contact between the reagents onto the flow and the catalyst (the enzyme) situated near the walls.

Considering the transesterification reaction, a crude vegetable oil, safflower oil from Carthamus tinctorius seed, has been chosen for transesterification with ethanol (Figure 2). This oil being principally composed of two unsaturated triglycerides, trilinolein (around 70 %) and triolein (around 10 %) completed with tripalmitin (8%).For this reaction we have previously demonstrated that the Candida rugosa lipase is not the candidate of choice, being less active than the Thermomyces Lanuginosus lipase[15]. Therefore, a new column has been synthesized using the same approach as previously and called Col[C-TL-lipase]@gGlymo-Si(HIPE). The evolution of the conversion in butyl linoleate versus time is reported in Figure 2. For a reagents solution flow rate of 0.05 mL min-1, the daily conversion level in butyl linoleate, initially close from 40%, decreases slowly but regularly to reach finally about 15% after two months of continuous conversion, this behavior can be attributed to both a consistent dehydration of the enzyme[16] and a degradation of the enzyme conformation through ethanol source. It is now possible to compare the performance of the Col[C-TL-lipase]@gGlymo-Si(HIPE) column with other supported-enzyme catalysis under continuous flux reported in the literature. Particularly it is reported that the use of lipase Candida sp confined within textile fibers is associated either with a 20 % decrease of enzymatic activity within 21 days of catalysis under continuous flux[17] or with a 16% decrease within 4 days of continuous catalytic reaction[10]. We have also to mention that a good enzymatic stability over 5 days (conversion of 80 % under continuous flow) has been obtained for stabilized lipase Candida antarctica B, (Novozyme 435), confined within macroporous acrylic resin in the transesterification of worn palm oil[18]. Using the same enzyme, Royan et al. have reported a conversion of 95 % during 21 days when considering the transesterification of coconut oilseed with methanol [19]. Under this study it has been established that tert-butanol, bearing a weak polarity, is favoring dissolution of glycerol by-products. In that case, 1.7 mg of enzyme per gram of oil has been used, allowing a methyl linoleate production of 0.17 μmol min^{-1} mg^{-1} (enzyme) during 21 days.

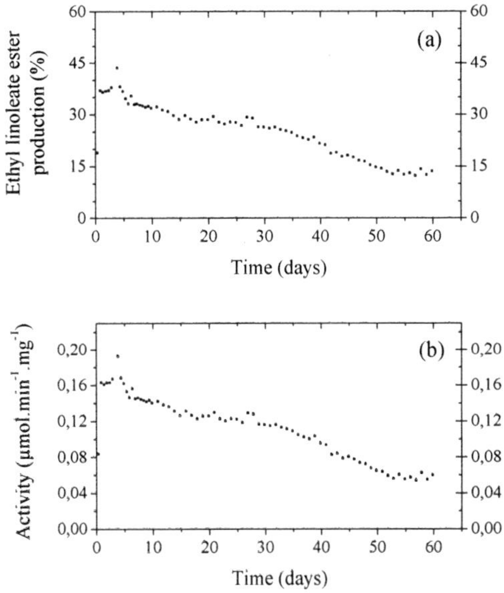

Figure 2. Catalytic performances for transesterification of safflower oil with ethanol in heptane obtained using Col[C-TL-lipase]@gGlymo-Si(HIPE) at 40°C at a 0.1 mL.min-1 flow rate. (a) Ethyl linoleate ester production versus time; (b) Enzymatic activity versus time. The enzymatic activity is expressed in µmol of ethyl linoleate ester produced per minute and per milligram of lipase immobilized.

In our case the methyl linoleate production starts at 0.16 µmol min^{-1} mg^{-1} (enzyme) with a small decrease to 0.13 µmol min^{-1} mg^{-1} (enzyme) after 21 days, production that is decreasing but still effective after two months of continuous flow catalysis at 0.06 µmol min-1 mg^{-1} (enzyme) (Figure 2b).

CONCLUSION

Herein we have proposed an original *in-situ* integrative chemistry [20-25] synthetic path dedicated to the one pot-synthesis and use of monolithic macroporous biohybrid catalysts. Such catalytic foams permit the design of continuous flow device with the advantages of covalent stabilization of the enzymes, low steric hindrance between proteins and substrates while offering an optimized mass transport through macroporous channels. Beyond the intrinsic simplicity in regard of the column *in-situ* synthetic path, these bio-hybrid heterogeneous catalysts provide among the top enzymatic activity addressed for catalyzed transesterification reactions bearing unprecedented endurance of continuous catalysis for a two months period of time. Despite this success, some improvements need to be reached as for instance performing solvent-free catalytic reactions while switching from food-dedicated oils toward wasted ones. These improvements are under study and will be published in due time.

REFERENCES

1. H. Fukuda, A. Kondo, H. Noda, *Journal of Bioscience and Bioengineering* **92**, 405 (2001).
2. M.I. Al-Widyan, A.O. Al-Shyoukh, *Bioresource Technology* **85**, 253 (2002).
3. M.J. Nye, T.W. Williamson, S. Deshpande, J.H. Shrader, W.H. Snively, T.P. Yurkewich, C.R. French, J. *Am. Oil Chem. Soc.* **60**, 1598 (1983).
4. M.S. Antczak, A. Kubiak, T. Antczak, S. Bielecki, *Renewable Energy* **34**, 1185 (2009).
5. A. M. Klibanov, *Science* **219**, 722 (1983).
6. G. Zhou, Y. Chen, S. Yan, *Micro. Meso. Mater.* **119**, 223 (2009).
7. C. F. Wang, G.W. Zhou, Y.J Li, N. Lu, H.B. Song, L. Zhang, *Colloids § Surfaces A: Physicochemical and Engineering Aspects* **406**, *75* (2012)
8. M. T. Reetz, A. Zonta, J. Simplekamp, *Biotechnology and Bioengineering* **49**, 527 (1996).
9. Z.J. Wang, M. Etienne, F Quiles, G.W., A. Walcarius, Biosens. Bioel. 32, 111 (2012).
10. Y. Chen, B. Xiao, J. Chang, Y. Fu, P. Lv, X. Wang, *Energy Conversion and Management* **50** 668 (2009).
11. P. Lozano, E. Garcia-Verdugo, R. Piamtongkam, N. Karbass, T. De Diego, M. I. Burguete, S. V. Luis, J.L. Iborra, *Adv. Synth. Catal.* **349**, 1077 (2007).
12. E. Calleri, C. Temporini, S. Furlanetto, F. Loiodice, G. Fracchiolla, G. Massolini, *Journal of Pharmaceutical and Biomedical Analysis* **32**, 715 (2003).
13. C. Mateo, V. Grazù, J.M. Palomo, F. Lopez-Gallego, R. Fernandez-Lafuente, J.M. Guisan, *Nature Protocol* **2**, 1022 (2007).
14. C. Mateo, G. Fernandez-Lorente, O. Abian, R. Fernandez-Lafuente, J.M. Guisan, *Biomacromolecule* **1**, 739 (2000).
15 . N. Brun, A. Babeau-Garcia, H. Deleuze, F. Duran, C. Sanchez, V. Ostreicher, R. Backov, Chem. Mater. **22**, 4555 (2010).
16. J.A. Rupley, G. Careri, *Adv. Protein Chem.* **41**, 37 (1991).
17. K. Nie, F. Xie, F. Wang,T. Tan, *J. Mol. Cat. B: Enzymatic* **43**, 142 (2006).
18. S.F.A. Halim, A. H. Kamaruddin, W.J.N. Fernando, *Bioresource Technology* **100**, 710 (2009).
19. D. Royon, M. daz, G. Ellenrieder, S. Locatelli, *Bioresource Technology* **98**, 648 (2007).
20. R. Backov, Soft Matter **2**, 452 (2006).
21. E. Prouzet, S. Ravaine, C. Sanchez, R. Backov, *New J. Chem.* **32**, 1284 (2008).
22. S. Ungureanu, M. Birot, L. Guillaumme, H. Deleuze, O. Babot, M.-F. Achard, M. I. Popa, C. Sanchez, R. Backov, *Colloids § Surfaces A: Physicochemical and Engineering Aspects* **360**, 85 (2010).
23. N. Brun, S. Ungureanu, H. Deleuze, R. Backov, *Chem. Soc. Rev.* **40**, 771 (2011).
24. N. Brun, B. Julian-Lopez, P. Hesemann, G. Laurent, H. Deleuze, C. Sanchez, M.-F. Achard, R. Backov, *Chem. Mater* **20**, 7117 (2008).
25. N Kinadjian, M. Le Bechec, T. Pigot, F. Dufour, A. Bentaleb, E. Prouzet, S. Lacombe, R. Backov, *Eur. J. Inorg. Chem.*, 5350 (2012).

Mater. Res. Soc. Symp. Proc. Vol. 1492 © 2013 Materials Research Society
DOI: 10.1557/opl.2013.219

Salt as Alternative Energy Material to Fossil Fuel
<potential use of sodium metal for hydrogen generation>

Masataka Murahara[1,2,3], Yuji Sato[4], and Toshio Okawara[3]
[1]Professor Emeritus of Tokai University, Hiratsuka, Kanagawa, Japan
[2]Innovative Research Initiatives, Tokyo Institute of Technology, Tokyo, Japan
[3]M Hikari & Energy Laboratory Co., Ltd., Kamakura, Kanagawa, Japan
[4]Institute for Laser Technology, Osaka, Japan

ABSTRACT

Salt is the raw material of sodium metal, which reacts with water to produce hydrogen for power generation. Sodium metal is solid matter and its specific gravity is low; therefore, it can be stored or transported for long at room temperature and under atmospheric pressure as oil and coal can. Sodium metal is produced with molten-salt electrolysis from sea salt, lake salt or rock salt, and securely kept immersed in kerosene for preventing it from reacting with air or moisture when transported to a consumer place; where it reacts violently with water to generate a large amount of hydrogen instantly. And sodium hydroxide, which is a reaction residue obtained after the production of hydrogen, is supplied as it is as the raw material of soda industries. Moreover, fresh water, sulfuric acid, hydrochloric acid, sodium hydroxide, and magnesium are generated as by-products in the processes of manufacturing sodium metal and generating hydrogen. Sodium metal can be an alternative energy material for hydrogen combustion power generation, having a far-reaching economic effect.

INTRODUCTION

Oil and coal have reigned in fossil fuel because they are kept well for long and transported for a long distance owing to their low specific gravities. Oil and coal are relatively light liquid and solid, respectively, and their combustion wastes can be disposed on site; which make their long-haul transportation and long-term storage possible [1]. Unfortunately, their reserve production ratios are limited as in Fig. 1. The reserve production ratio of fossil fuel is as follows [2]: 41 years for oil, 67 years for natural gas, and 165 years for coal. Nuclear power generation has been reviewed as a fuel that does not release CO_2, but uranium is also exhaustive, 85 years in reserve production ratio. The depletion of the fuel resources is concerned, and the current procurement of energy sources is limited to places rich in resources. However, salt, which is the raw material of sodium metal, is present inexhaustibly in sea water and also exists as a rock salt or salt lake on the land abundantly. It is not necessary to worry about maldistribution and exhaustion of resources.

Sodium metal, which is produced with molten-salt electrolysis, generates hydrogen when adding water. Hydrogen is a very light gas, but a gas cylinder for its storage is very heavy and unsuitable for transportation. Being a solid material with a low specific gravity, sodium metal can be stored for long and transported to a consumer place in the distance at room temperature and under atmospheric pressure like oil and coal. At a thermoelectric power station in the consumer place transported, it reacts violently with water to generate hydrogen for hydrogen-fueled combustion power generation. Hydrogen does not discharge CO_2, being a clean and environmentally friendly fuel, although oil and coal discharge CO_2 when burnt.

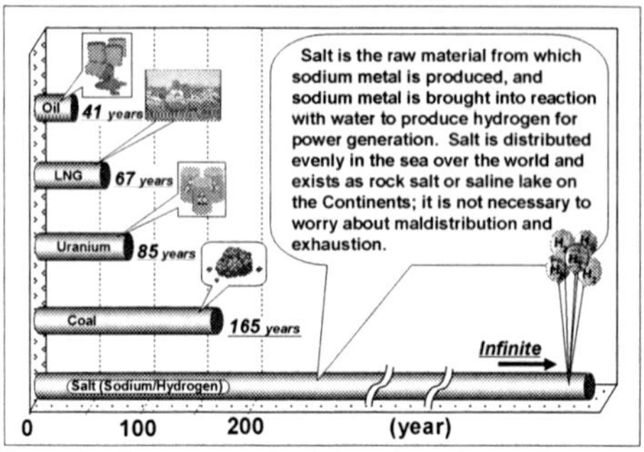

Figure 1. Reserve production ratios of natural resources.

The generation of electricity by natural energy such as wind power or solar power is influenced by climate conditions, so the net generation is not constant. The uncertainness of the weather can be disregarded if the natural energy is converted to a solid fuel, and it will make the long-term storage and long-distance transportation possible. It is considered here to use sodium metal that is solid matter as an energy material. Sodium metal is produced with molten-salt electrolysis and is securely kept immersed in kerosene to transport to a consumer place; where sodium metal and water react each other to generate hydrogen instantly for power generation [3]. Sodium hydroxide, a reaction residue obtained after the production of hydrogen, is usually a starting material of soda industries. And, sodium metal generally needs high power for production, which is very expensive. In the processes of manufacturing sodium metal and generating hydrogen, however, fresh water, sulfuric acid, hydrochloric acid, sodium hydroxide, and magnesium are generated as by-products. These materials that have also consumed a large amount of electricity for production are produced without any cost. The economical efficiency is high. Moreover, the production factory is conveniently located on the raw material, seawater or rock salt, and the power plant is operated using natural energy such as wind power or photovoltaic power. The economic benefit, therefore, is enormous overall [1], [3], [4], [5].

In offshore wind power generation, it is expected that wind turbines be installed at wind farms in the open sea free from the fishery right in the future, which will increase power transmission loss and installation cost of power transmission line further. The problem is how to transport the energy resources economically to a consumer place. To solve the problem, it requires a temporary storage; which is again sodium metal.

THEORY

Salt is dissolved in enormous sea water on the ocean

The contents of the elements dissolved in seawater differ more or less with the conditions such as a water temperature, salinity, and the depth of a collection place, but there are no striking places rich in elements. It is not necessary to follow the same procedure of prospecting, mining, and mineral dressing as the one for coal, oil, or mine development, but it is just pumping and refining the seawater that contains minerals. And it is relatively easy to collect an object substance from the seawater by electrolytic, chemical, or thermal analysis.

As shown in Fig. 2, seawater consists of water 96%, salt 3.5%, and other metallic salts the rest. The 3.5% salt contains calcite ($CaCO_3$), gypsum ($CaSO_4$), sodium chloride ($NaCl$), magnesium sulfate ($MgSO_4$), magnesium chloride ($MgCl_2$), and potassium chloride (KCl). NaCl that is contained most in the seawater is electrolyzed to produce sodium metal [5].

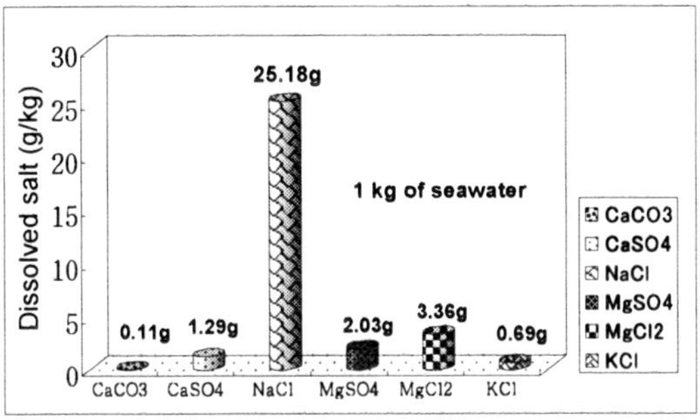

Figure 2. Crystallization of salt from seawater 1 kg. When the seawater is put on a plate and heated, the water evaporates gradually, and the substances start precipitating in order of solubility from low to high.

Rock salt and salt lake exist on the land

Rock salt and salt lakes are present on the Continents, as in Fig. 3. There are many wind farms, rock salt areas, and salt lakes in North America and Europe; they exist in Bolivia, Argentina, Columbia, Ecuador, Peru, and Chile in South America, and also in China, Ukraine, Pakistan, Iran, and Mongolia in South West Asia. Patagonia in the Argentine and Sinkiang

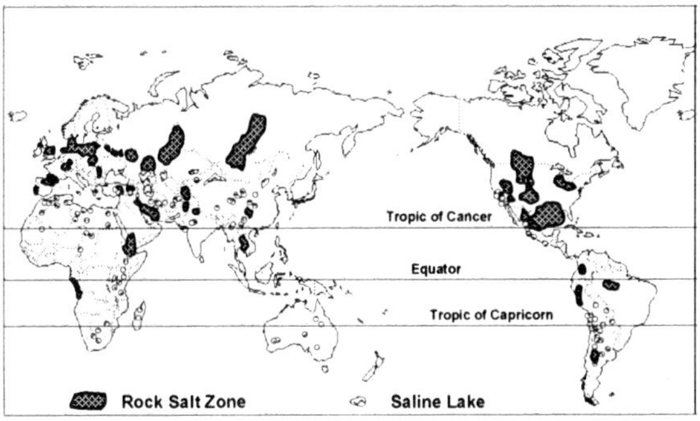

Figure 3. Distribution chart of rock salt and salt lakes on the Continents.

Uighur in China are famous for strong wind and have salt lakes. Many rock salt deposits are found to date. The reserves is almost unlimited, and the rock salt currently occupies three quarters of the amount of the salts produced in the world.

Figure 4 shows the photos taken at Uyuni Salt Flats located in the southwest of Bolivia, which is the world's largest salt flats at 3,656 meters above sea level and whose area is 10,582 square kilometers. It is covered with the salt crust of a few meters. The crust serves as a source of salt (a) and as a cover of the pool of brine (b), which contains lithium 10,000 times and sodium 6.5 times as much as sea water does [6]. Sodium metal may be produced in lakes and rock salt drilling sites in these areas and transported to consumer places. A production system in such an area having raw materials and power sources jointly can reduce an energy loss in production, storage, and transportation and enables the provision of an energy production plant capable of improving the efficiency of the whole system [1], [4].

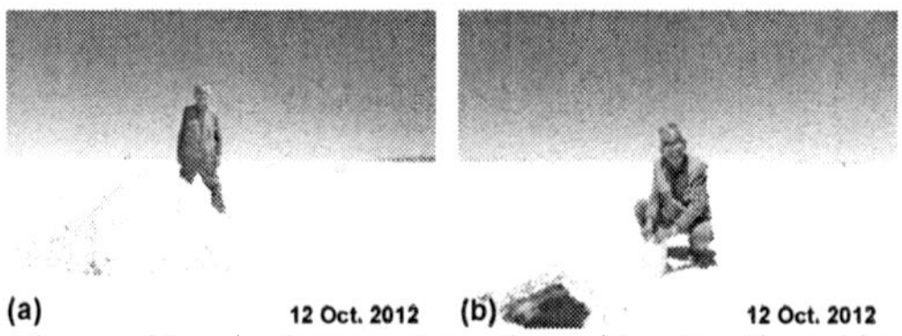

Figure 4. Photos taken on Uyuni salt flats: (a) author with salt collected by scratching and (b) author collecting brine underneath the salt crust.

On-site sodium metal production for hydrogen generation on ocean and on-demand hydrogen power generation on land

The conceptual diagram of sodium metal production for hydrogen generation by using seawater and offshore wind power generation is shown in Fig. 5. The on-site electrolytic plant is located on the raw material called "seawater" and is operated by natural energy such as wind power, hydraulic power, sunlight, or solar heat. Utilizing the power and seawater, sodium metal is produced on site by molten-salt electrolysis and transported to consumer places on land, where

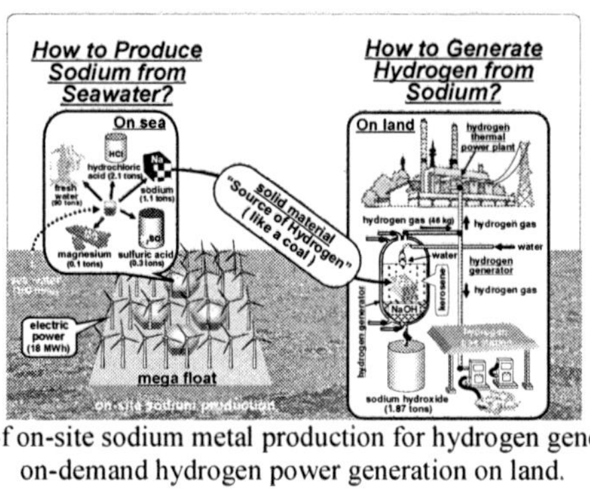

Figure 5. Diagram of on-site sodium metal production for hydrogen generation on ocean and on-demand hydrogen power generation on land.

it is brought into reaction with water for hydrogen generation.

Generally, sodium metal production requires high power, which is very expensive. In the processes of manufacturing sodium metal and generating hydrogen, however, fresh water, sulfuric acid, hydrochloric acid, sodium hydroxide, and magnesium are generated as by-products. These by-products also consume high power when producing, but by this production system, they are produced without any cost. And, the sodium metal is transported to thermal power plants in consumer places on land, where water and sodium metal are reacted with each other to generate hydrogen as in the following equation ($Na + H_2O \rightarrow 1/2H_2 + NaOH$) for hydrogen combustion power generation. Sodium hydroxide, which is a reaction residue obtained after the production of hydrogen, is supplied as it is as the raw material of soda industries. The system is, therefore, considered to have an enormous economic effect [4], [5].

DISCUSSION

An amount of power required at the offshore electrolytic plant is estimated. Recovery systems and devices would be improved in the future, and so would recovery efficiency; thus, the estimation is made on the assumption that without setting recovery efficiency, all the elements dissolved in seawater will be collected. As in Fig. 6, assuming that the plant treats 10^5 tons of seawater, 17.8 GW/day is consumed to produce 9×10^4 tons of fresh water, 1,075 tons of sodium metal, 129 tons of magnesium, 278 tons of sulfuric acid, and 2,091 tons of hydrochloric acid. The 1,075 tons of sodium metal is transported to the land, where it is brought into reaction with 843.3 tons of water; the amounts of hydrogen and sodium hydroxide generated are 46.7 tons and 1,867 tons, respectively [1], [5]. The amount of hydrogen produced by electrolyzing water directly is 317 tons and approximately six times as large as that of hydrogen produced by molten-salt electrolysis (46.7 tons), but no by-products are produced in the manufacturing process. Therefore, the final production cost increases by five times [1], [5].

The market price of electricity is US$0.1/kWh, on the average. Those of hydrogen, fresh water, sulfuric acid, magnesium metal, sodium hydroxide, and hydrochloric acid are US$2,000, $1, $700, $4000, $600, and $500 per ton, respectively. When the operating rate of wind power generation is 42% or 10 hours per day, $1.78 million (1.78 GWh × 10 × $0.1/kWh) of electricity is consumed to treat the 0.1 megatons of seawater and produce 1,075 tons of sodium

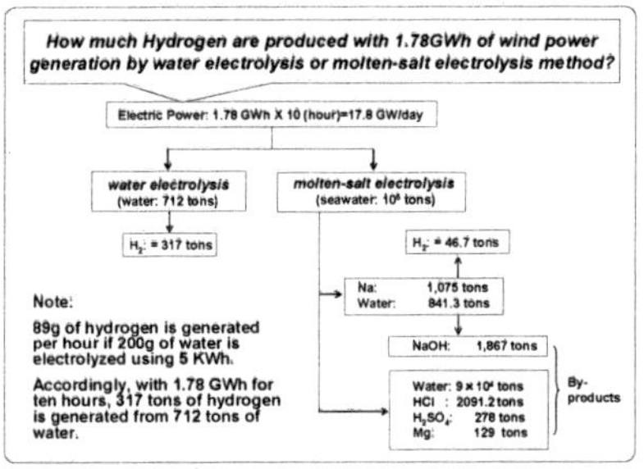

Figure 6. Estimates of outputs from 10^5 tons of seawater by power consumption of 17.8 GW/day.

metal, the object product; from which only \$93.4 thousand (46.7 tons × \$2,000/ton) of hydrogen, the main target, is generated. The yield is very low. However, many by-products are produced in the manufacturing process: \$0.09 million (90 kilotons × \$1/ton) of fresh water, \$0.195 million (278 tons × \$700/ton) of sulfuric acid, \$0.516 million (129 tons × \$4,000/ton) of magnesium metal, \$1.12 million (1,867 tons × \$600/ton) of sodium hydroxide, and \$1.046 million (2091.2 tons × \$500/ton) of hydrochloric acid. The sum of the by-products is extraordinarily high, \$2.97 million. Consequently, the total output is \$3.06 million, thanks to many by-products. Production efficiency is calculated at 172% (\$3.06 million/\$1.78 million). Although it is expensive, the sodium metal production system is still profitable because many by-products of great value are yielded without cost [1], [5].

CONCLUSIONS

Sodium metal is present inexhaustibly as salts in sea water and also exists as a rock salt or salt lake on the land abundantly. It is not necessary to worry about maldistribution and exhaustion of resources. It not only solves the maldistribution problem of fossil fuels [3],[5], which is one of the factors to threaten the energy security, but also minimizes the damages or disorders caused by natural disasters or terrorist attacks if occurred by distributing sodium metal produced on-site to thermal power plants in the stricken areas for power generation. Besides, the hydrogen-fueled combustion turbines emit nothing but water, no CO_2 or radioactivity at all. It contributes to the environmental protection of the earth as an environmentally friendly power plant.

Utilizing these chemicals while maintaining the modern industrial structure enables the world to be environmentally friendly with renewable energy and without resources wars, which ensures the stable supply of water, energy, and industrial chemicals [4],[5]. There is no fuel except sodium metal that discharges neither CO_2 nor radioactivity and is depletion-free [1], [5].

REFERENCES

1. M. Murahara and K. Seki, On-Site Electrolysis Sodium Metal Production by Offshore Wind or Solar Energy for Hydrogen Storage and Hydrogen Fuel Cycle, *Proc. IEEE/ECCE 2010 Conf. ENERGY CONVERSION CONGRESS & EXPO*, Atlanta, GE, 2010, 4264-4269. (IEEE, ISBN: 978-1-4244-5287-3/10)
2. M. S. Dresselhaus and I. L. Thomas, Alternative energy technologies, *Nature, 414(15)*, 2001,332-337.
3. M. Murahara and et al., Editor: Ruth A. Reck, *Climate Change and Sustainable Development*, (Cambridge, UK, Chicago, New York: Linton Atlantic Books, Ltd., 2010) 215-221.
4. M. Murahara, On-site integrated production plant, International Patent Application No. PCT/JP2008/ 058500, 2008, (International Disclosure Number: WO2008-142995), China Patent: No.876024, European patent: No.2149625, USA patent: No.8197664, No. 8277632
5. M. Murahara, Renewable Energy --- *Last Hurrah for Nuclear Power Plant*, (Tokyo: Power-Sha, 2011) (in Japanese).
6. "Uyuni Salt Flats", Encyclopedia Britannica. Retrieved 2007-12-01

AUTHOR INDEX

SUBJECT INDEX

CPSIA information can be obtained at www.ICGtesting.com
Printed in the USA
LVOW10*0324250114

370927LV00001B/1/P

9 781605 114699